W9-BKK-656

# METALOGIC

## *An Introduction to the Metatheory of Standard First Order Logic*

Geoffrey Hunter

*Senior Lecturer in the Department of Logic and Metaphysics*
*University of St Andrews*

UNIVERSITY OF CALIFORNIA PRESS
Berkeley and Los Angeles 1971

UNIVERSITY OF CALIFORNIA PRESS
Berkeley and Los Angeles, California

ISBN: 0-520-01822-2
Library of Congress Catalog Card
Number: 71-131195



*Printed in Great Britain*

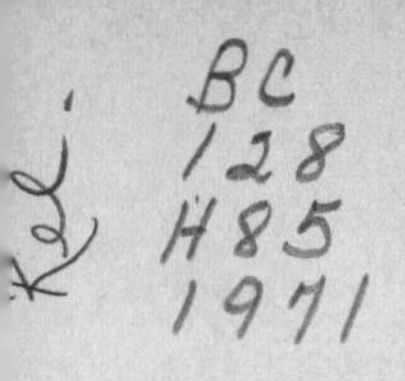

To my mother and to the memory of my father,

Joseph Walter Hunter

# Contents

## *Part Three*: First Order Predicate Logic: Consistency and Completeness

# Preface

My main aim is to make accessible to readers without any specialist training in mathematics, and with only an elementary knowledge of modern logic, complete proofs of the fundamental metatheorems of standard (i.e. basically truth-functional) first order logic, including a complete proof of the undecidability of a system of first order predicate logic with identity.

Many elementary logic books stop just where the subject gets interesting. This book starts at that point and goes through the interesting parts, as far as and including a proof that

> it is impossible to program a computer to give the right answer (and no wrong answer) to each question of the form 'Is — a truth of pure logic?'

The book is intended for non-mathematicians, and concepts of mathematics and set theory are explained as they are needed.

The main contents are: Proofs of the consistency, completeness and decidability of a formal system of standard truth-functional propositional logic. The same for first order monadic predicate logic. Proofs of the consistency and completeness of a formal system of first order predicate logic. Proofs of the consistency, completeness and undecidability of a formal system of first order predicate logic with identity. A proof of the existence of a non-standard model of a formal system of arithmetic.

The reader will be assumed to have an elementary knowledge of truth-functional connectives, truth tables and quantifiers. For the reader with no knowledge of set theory, here are very brief explanations of some notations and ideas that will be taken for granted later on:

1. *The curly bracket notation for sets*

'{Fido, Joe}' means 'The set whose sole members are Fido and Joe'. '{3, 2, 1, 3, 2}' means 'The set whose sole members are the numbers 3, 2, 1, 3, 2' (and this last set is the same set as {1, 2, 3}, i.e. the set whose sole members are the numbers 1, 2 and 3).

2. *The epsilon notation for set-membership*

'$n \in X$' means '$n$ is a member of the set X'.

3. *The criterion of identity for sets*

A set A is the same set as a set B if and only if A and B have exactly the same members. *Nothing else matters* for set identity.

4. *The empty set*, $\emptyset$

By the criterion of identity for sets [(3) above], if A is a set with no members and B is a set with no members, then A is the same set as B; so if there is a set with no members, there is just one such set. We shall assume that there is such a set.

Further introductory material on set theory can be found in, for example, chap. 9 of Suppes (1957) or chap. 1 of Fraenkel (1961).

The book deals only with (1) *standard* (i.e. basically truth-functional) logic, and (2) *axiomatic* systems.

(1) Standard first order logic, with its metatheory, is now a secure field of knowledge; it is not the whole of logic, but it is important, and it is a jumping-off point for most other developments in modern logic. There seemed to me to be no book that tried to make accessible to non-mathematicians complete proofs of the basic metatheory of standard logic: hence this one.

(2) Axiomless systems (so-called 'natural deduction systems') are nowadays getting more popular than axiomatic systems, for formal proofs of theorems *inside* a system are generally shorter and easier to find with a natural deduction system than with an axiomatic one. But I find that complete proofs of *metatheorems* (theorems *about* a system) are in general longer and more laborious for natural deduction systems than for axiomatic ones; so, since I am mainly concerned with proofs of theorems about systems and not much concerned with proofs inside systems, and since anything you can get with a natural deduction system you can get with an axiomatic one, I have deliberately concentrated on axiomatic systems.

I hope that the book will equip those who are not mathematical specialists not only to tackle more advanced works on standard logic, such as those of Kleene or Mendelson or Shoenfield or Smullyan,[1] but to frame philosophically interesting systems of non-standard logic and to prove metatheorems about them. For I believe that the logician's most urgent tasks at present lie in the field of non-standard logic. There is, for example, a sense of 'if' that is crucial to many everyday arguments; it seems clear to me that 'if' in that sense is not a truth-functional connective; and so I think it a scandal that the sense has not yet been adequately caught in any interpreted formal system with an adequate metatheory. I commend the task to my readers, who may find help in Church (1956) and in Hughes and Cresswell (1968).

The books I have borrowed most from are Mendelson (1964) and Margaris (1967). I thank the following people for ideas, information, or criticisms: Ross Brady, John Crossley, John Derrick, Len Goddard, Jeff Graves, Geoffrey Keene, Martin Löb, Angus Macintyre, Timothy Potts, Rowan Rockingham Gill, Harold Simmons, Dick Smith and Bob Stoothoff. My predecessors and colleagues in the Department of Logic and Metaphysics in the University of St Andrews, and especially Professors J. N. Wright and L. Goddard, made it possible for me to concentrate on this part of logic, and without them the book would not exist.

GEOFFREY HUNTER

*St Andrews,*
*December 1969*

[1] For detailed references, see pp. 262 ff.

PART ONE

# Introduction: General Notions

To get a rough idea of what the metatheory of logic is, start with

(1) *truths of logic.*

Distinguish these from

(2) *sentences used to express truths of logic.*

(Two different sentences, e.g. one in French, one in English, might be used to express the same truth of logic.)

Now consider

(3) *the theory of sentences-used-to-express-truths-of-logic.*

This last is, roughly, *the metatheory of logic.*

The big difference between metatheory in that rough sense and metatheory in the sense of this book is over (2). In this book the sentences-used-to-express-truths-of-logic must be formulas of a *formal language*, i.e. a 'language' that can be completely specified without any reference at all, direct or indirect, to the meaning of the formulas of the 'language'. It was by the insistence on this requirement that the metatheory of logic, after a long and interesting but desultory history of over two thousand years, came in this century to yield exact and new and deep results and to give promise of systematic growth.

So we begin with formal languages.

## 1 Formal languages

The basic objects of metatheory are *formal languages*.

The essential thing about a formal language is that, even if it is given an interpretation, *it can be completely defined without reference to any interpretation for it*: and it need not be given any interpretation.

A formal language can be identified with the set of its *well-formed formulas* (also called *formulas* or *wffs*). If the set of all wffs of a formal language L is exactly the same as the set of all wffs of a formal language L′, then L is the same formal language as L′. If not, not.

A *formula* is an abstract thing. A *token* of a formula is a mark or a string of marks. Two different strings of marks may be tokens of the same formula. It is not necessary for the existence of a formula that there should be any tokens of it. (We want, for example, to speak of formal languages with infinitely many formulas.)

The set of well-formed formulas of a particular formal language is determined by a fiat of its creator, who simply lays down what things are to be wffs of his language. Usually he does this by specifying

(1) a set of *symbols* (the *alphabet* of his language)

and

(2) a set of *formation rules* determining which sequences of symbols from his alphabet are wffs of his language.

It must be possible to define both sets without any reference to interpretation: otherwise the language is not a formal language.

The word 'symbols' in the last paragraph is a technical term: symbols, in this technical sense of the word, need not be symbols of anything, and they must be capable of being specified without reference to any interpretation for them.

Symbols are abstract things, like formulas. A token of a symbol is a mark or configuration of marks.

Roughly, a formal language could be completely mastered by a suitable machine, without any understanding. (This needs qualification where the formal language has an uncountable alphabet (see §10 below): in such a case it is not clear that the formal language could be completely mastered by anything.)

Given a particular formal language, we may go on to do either or both of the following things:

1. We may define the notion of *an interpretation of the language*. This takes us into *model theory*.

2. We may specify a *deductive apparatus* for the language. This takes us into *proof theory*.

EXERCISES

1. The language W is defined as follows:
   Alphabet: △ □
   Formulas: Any finite string of symbols from the alphabet of W that begins with a '△' is a formula.
   Is W a formal language?
2. The language X is defined as follows:
   Alphabet: a b c d e f g
   Formulas: Any finite string of symbols from the alphabet of X that makes an English word is a formula.
   Is X a formal language?
3. The language Y is defined as follows:
   Alphabet: a b c d e f g
   Formulas: Any finite string of symbols from the alphabet of Y that does not make an English word is a formula.
   Is Y a formal language?

ANSWERS

1. Yes.

2. No. The definition of *formula of X* involves essentially reference to meaning, since a thing is an English word only if it has a meaning. (In order to know that a thing is a word you don't have to know *what* it means, only *that* it has a meaning. But even this weak reference to meaning is enough to prevent X from being a formal language.) (Another way of putting it: You could program a machine to find out if the string of symbols was a word in some specified English dictionary, and the

machine could do this without knowing the meaning of any word. But, with some exceptions which we may neglect, things are included in a dictionary only if they have a meaning or meanings.)

3. No. In order to tell whether or not a string of symbols from the alphabet of Y is a formula of Y you have to know whether or not it is an English word, and so whether or not it has a meaning. How, for example, do you tell whether 'bac', or 'deg', or 'ged', or 'gef', or 'geg', or 'gegg', is a formula of Y? Only by finding out if it is a meaningful English word. (In fact 'geg' seems to be the only one that is not an English word, and so the only one that is a formula of Y.)

## 2 Interpretations of formal languages. Model theory

In rough and very general terms, an *interpretation* of a formal language is an assignment of meanings to its symbols and/or formulas.[1] *Model theory* is the theory of interpretations of formal languages (a *model* of a formula of a language is an interpretation of the language for which the formula comes out true).[2] Among the concepts of model theory are those of *truth for an interpretation*, *semantic* (or *model-theoretic*) *consequence*, and *logical validity*.

EXERCISE

Give an interpretation for the formal language W (§1, exercise 1).

ANSWER

A possible interpretation would be: Take '△' as meaning the same as the (decimal) digit '1', '□' as meaning the same as the digit '0', and each formula accordingly as meaning the same as

[1] In Part 2 we restrict the notion of interpretation to interpretations for which each interpreted formula is either true or false. In Part 3 there is an analogous restriction.

[2] At least in one standard sense of 'model', which we follow in this book. Another sense, closely related to ours, will be mentioned in Part 3.

a decimal numeral composed exclusively of '1's and '0's. So, e.g., '△ □ △' is to mean the same as '1 0 1' in the decimal system.

This shows that, in a very wide sense of 'interpretation', an interpreted formula need not be a proposition, where by 'proposition' we mean a sentence expressing something true or false. It can be, as here, a name of something. Or it can be an adjective, or an adverb, or a preposition, or a phrase, or a clause, or an imperative sentence, or a string of sentences, or a string of names, or . . . Or such meanings might be attached to the symbols that some or all of the interpreted formulas came out as nonsense. Later in the book we restrict the notion of interpretation: see p. 6, fn. 1.

## 3 Deductive apparatuses. Formal systems. Proof theory

By specifying a deductive apparatus for a formal language we get a formal system.

A *formal system* S is a formal language L together with a *deductive apparatus* given by

(1) laying down by fiat that certain formulas of L are to be *axioms* of S

and/or

(2) laying down by fiat a set of *transformation rules* (also called *rules of inference*) that determines which relations between formulas of L are relations of *immediate consequence* in S. (Intuitively, the transformation rules license the derivation of some formulas from others.)

The deductive apparatus must be definable without reference to any intended interpretation of the language: otherwise the system is not a formal system.

A deductive apparatus can consist of axioms and rules of inference, or of axioms alone, or of rules of inference alone.

*Proof theory* is that part of the theory of formal systems (i.e. of formal languages with deductive apparatuses) that does not involve model theory in an essential way (i.e. that does not require any reference to interpretations of the languages). Among concepts belonging to proof theory are those of *proof in a system* (or *formal proof*), *theorem of a system* (or *formal theorem*), *derivation in a system* (or *formal derivation*), and *syntactic* (or *proof-theoretic*) *consequence*. All these involve essentially reference to a deductive apparatus, and all can be defined without saying anything about interpretations.

A formal language can be identified with the set of all its wffs. But a formal system cannot be identified with the set of all its theorems. For two formal systems S and S′ may have exactly the same theorems and yet differ in some proof-theoretically important way: e.g. a formula A that is a syntactic consequence in S of a formula B may not be a syntactic consequence in S′ of B.

## EXERCISES

Let Z be the system defined as follows:

Alphabet: △ □

Formulas: Any finite string of symbols from the alphabet of Z that begins with a '△' is a formula of Z. Nothing else is a formula of Z.

Axiom: △ □ □ □

Rule of Inference: Any formula of Z whose last two symbols are a '△' and a '□', in that order, is an immediate consequence in Z of any formula of Z whose first two symbols are a '△' and a '□', in that order. [E.g. '△ □ □ □ △ △ □' is an immediate consequence in Z of '△ □ △ △ △ □ △'.] Nothing else is an immediate consequence in Z of anything.

1. Is Z a formal system?
2. Is '△ △ □' an immediate consequence in Z of '△ □ □ □'?
3. Is '△ □' an immediate consequence in Z of '△ □'?
4. Is '△ □ □ △' an immediate consequence in Z of '△ □ □ △'?
5. Is '□ □ △ △ □' an immediate consequence in Z of '△ □ △ △ △ △'?
6. Give an example of an immediate consequence in Z of '△ △ △ △ △'.

ANSWERS

1. Yes.
2. Yes.
3. Yes.
4. No. Only formulas that end '. . . △ □' can be immediate consequences in Z.
5. No. '□ □ △ △ □' is not a formula of Z, since it does not begin with a '△'.
6. There are none. Only formulas that begin '△ □ . . .' can have immediate consequences in Z.

## 4 'Syntactic', 'Semantic'

In this book 'syntactic' and 'semantic' will have the following meanings:

*Syntactic*: having to do with formal languages or formal systems without essential regard to their interpretation.

*Semantic*: having to do with the interpretation of formal languages.

'Syntactic' has a slightly wider sense than 'proof-theoretic', since it can be applied to properties of formal languages without deductive apparatuses, as well as to properties of formal systems. 'Semantic' in this book just means 'model-theoretic'.

EXERCISES

1. Is it a syntactic or a semantic property of a formula of the system Z (§3, exercises) that it is an immediate consequence in Z of another formula of Z?
2. Is it a syntactic or a semantic property of a formula that it denotes a number?
3. Is it a syntactic or a semantic property of a formula that it is true?

ANSWERS

1. Syntactic.
2. Semantic: the last part of the question can be paraphrased by '. . . that it can be *interpreted* as denoting a number'.

3. Semantic: the last part of the question can be paraphrased by '. . . that it can be *interpreted* as expressing something true'.

## 5 Metatheory. The metatheory of logic

*Metatheory* is the theory of formal languages and systems and their interpretations. It takes formal languages and systems and their interpretations as its objects of study, and consists in a body of truths and conjectures about these objects. Among its main problems are problems about the *consistency*, *completeness* (in various senses), *decidability* (see §8 below) and *independence* of sets of formulas. Both model theory and proof theory belong to metatheory.

The *metatheory of logic* is the theory of those formal languages and systems that for one reason or another matter to the logician. Usually the logician is interested in a formal language because it has formulas that can be interpreted as expressing logical truths; and usually he is interested in a formal system because its theorems can be interpreted as expressing logical truths or because its transformation rules can be interpreted as logically valid rules of inference.

## 6 Using and mentioning. Object language and metalanguage. Proofs in a formal system and proofs about a formal system. Theorem and metatheorem

In logic the words 'use' and 'mention' [both the nouns and the verbs] are sometimes used in a technical sense to mark an important distinction, which we explain by example:

A. 'London' is a word six letters long.
B. London is a city.

In A the word 'London' is said to be *mentioned*; in B the word 'London' is said to be *used* (and not mentioned).

There are various ways of indicating that an expression is being mentioned; e.g. enclosing it in quotation marks (as in the example), or printing it in italics, or printing it on a line to itself. We shall use some of these devices, but in order to save quotation marks and also to make the text easier to read, we shall in addition make use of a standard convention. In cases where the context makes it clear that expressions are being mentioned, not used, we shall sometimes omit quotation marks: e.g. instead of writing

'$\supset$' is a truth-functional connective

we shall write simply

$\supset$ is a truth-functional connective

and instead of writing

The set {'$\sim$', '$\wedge$', '$\vee$'}

we shall write simply

The set $\{\sim, \wedge, \vee\}$.

Formal languages are sometimes called *object languages*. The language used to describe an object language is called its *metalanguage*. We use English, supplemented by special symbolism (including the symbolism of set theory), for our metalanguage.

A *proof in a formal system* that has axioms (and all the ones we shall be concerned with have axioms) is a string of formulas of a formal language that satisfies certain purely syntactic requirements and has no meaning.

A *proof about a formal system* is a piece of meaningful discourse, expressed in the metalanguage, justifying a true statement about the system.

Similarly a *theorem of a formal system* is a formula of a formal language that satisfies certain purely syntactic requirements and has no meaning, while a *theorem about a formal system* (also called a *metatheorem*) is a true statement about the system, expressed in the metalanguage.

EXERCISES

1. *Christian lady*: It is enough for my argument if you admit that the existence of God, if not certain, is at least probable; or if not probable, is at least possible.

*Infidel*: I can make no such admission until I know what you intend by the word 'God'.

Charles Bradlaugh, 'Doubts in Dialogue', *National Reformer*, 23 Jan 1887

(*a*) In the dialogue above, is the Christian lady using or mentioning the word 'God', in the logician's special sense of the words 'using' and 'mentioning'?

(*b*) Is the Infidel using or mentioning the word 'God'?

2. In each of the following cases say whether the sentence 'Shut the door' is used or mentioned:

(*a*) 'Shut the door' is used to make a request or issue a command.

(*b*) Shut the door.

(*c*) I don't know what you mean by 'Shut the door'.

3. In each of the following cases say whether the sentence 'Cruelty is wrong' is used or mentioned:

(*a*) Cruelty is wrong, no matter what the circumstances.

(*b*) The words *Cruelty is wrong* express a true proposition.

(*c*) The sentence 'Cruelty is wrong' is typically used to condemn cruelty.

4. In the following sentence which words are used, which mentioned?

> What *is* means is and therefore differs from *is*, for '*is* is' would be nonsense.

Bertrand Russell, *My Philosophical Development*, p. 63

5. Is the proposition

> Not every string of '△'s and '□'s is a formula of the formal system Z [of §3]

a theorem of Z, a theorem about Z, or a metatheorem?

6. Given that, for any formal system S, anything that is either an axiom of S or an immediate consequence in S of an axiom of S is a *theorem* of S, say for each of the following whether or not it is a theorem of the system Z of §3. [*Note*. This is not intended to be a definition of the notion of *formal theorem*, but merely a simplification for the sake of the exercise. The usual definition allows many more things to be theorems.]

(*a*) △ □ △ □

(*b*) □ □ △ □ [For System Z, see p. 8.]

(*c*) △ □ □ □

(*d*) '△ □ □ □' is a theorem of Z.

7. 'Let A and B be arbitrary formulas of a formal language

L.' Explain the function in this sentence of the letters 'A' and 'B'.

8. In the exercises on §3 are the quotation marks in all and only those places where they ought to be?

ANSWERS

1. (*a*) Using.
   (*b*) Mentioning.
2. (*a*) Mentioned.
   (*b*) Used.
   (*c*) Mentioned.
3. (*a*) Used.
   (*b*) Mentioned.
   (*c*) Mentioned.

4. The sentence is 15 words long. I think that the 2nd, 9th, 11th and 12th words are mentioned, the others used. The 2nd and 9th words are names of the word 'is'. The first of the two words in quotation marks is the name of the word 'is'; the second is the word 'is'. So at that point first the name of the word 'is' is mentioned, and then the word 'is' is mentioned.

5. It is both a theorem about Z and a metatheorem (about Z). It is not a formula of Z, so it is not a theorem of Z.

6. (*a*) Yes (immediate consequence of the axiom).
   (*b*) No (not even a formula – it begins with a '□').
   (*c*) Yes (axiom).
   (*d*) No. Not a formula of Z (no formula of any of the formal languages in this book contains an expression in quotation marks). But it is a metatheorem of Z.

7. 'A' and 'B' here are metalinguistic variables, belonging to the metalanguage of the language L.

8. I hope so.

## 7 The notion of *effective method* in logic and mathematics

Throughout what follows we are concerned with effective methods *in logic and mathematics*, and not, e.g., with methods for telling whether or not something is an acid.

In logic and mathematics, an *effective method* for solving a problem is a method for computing the answer that, if followed correctly and as far as may be necessary, is logically bound to give the right answer (and no wrong answers) in a finite number of steps. An effective method for the solution of a class of problems is an effective method that works for each problem in the class.

This is not a very precise definition, but the concept is not a precise one, even though it belongs to the fields of logic, mathematics and computing.

The paradigm cases of effective methods are mathematical algorithms, such as Euclid's algorithm (Book VII, Prop. 1) for telling whether or not two positive integers have any common divisor other than 1, or his algorithm (Book VII, Prop. 2) for finding the greatest common divisor of two positive integers that are not relatively prime (i.e. that have some common divisor other than 1).

A method can be effective even though it is not possible in practice to follow it as far as is necessary in some (or even any) given case. For instance, there are numbers so large that writing or printing out their names in any suitable notation would take more paper than there is in the world; so (barring quite extraordinary powers of mental arithmetic) it would not be possible in practice to find their greatest common divisor by means of Euclid's algorithm. Nevertheless, even for those numbers Euclid's algorithm is an effective method.

It is not necessary for the existence of an effective method that it should be known to someone at some time. There may exist effective methods that nobody ever discovers in the whole history of the world.

Because an effective method must be capable of being followed mechanically, without requiring any insight or imagination or ingenuity on the part of the user, 'mechanical method' is sometimes used as a synonym for 'effective method'. Our definition ensures that an effective method must not require imagination or ingenuity by making it a condition of a thing's being an effective method that (1) it is a method of *computing* and (2) if it is followed correctly and as far as may be necessary, it is *logically* bound to give the right answer.

An objection to our explanation of 'effective method' is that we have left unexplained the notions of *computing* and *logically bound* that are crucial to it. Our present answer to this objection is that the notion of an effective method is an informal, intuitive, imprecise one, not a formal and precise one, so that an explanation of the intuitive sense is bound to be imprecise. In a later Part we mention some suggested definitions of effectiveness that are precise and for which it is claimed that they do correspond satisfactorily to the intuitive notion (§52, Church's Thesis).

EXERCISES

1. Is 'Ask God' an effective method for solving a problem?
2. Is 'Ask an oracle' an effective method?
3. Is 'Ask an oracle that always answers and always tells the truth' an effective method?
4. Is 'First say "Yes", then say "No"' an effective method for solving a problem to which the right answer happens in fact to be 'Yes'?
5. Is 'Test it with litmus paper' an effective method (in the sense defined) for telling whether or not something is an acid?
6. 'No solution has been found to this problem, so there is no effective method for solving it.' Is this a valid argument?

ANSWERS

1. No. For (*a*) it is not a method of *computing*, and (*b*) it is not logically bound to give the right answer, for God need not answer.
2. No. Reasons as for 1 above.
3. No. Not a method of *computing*.
4. No. Cf. the requirement in the definition that the method should give *no wrong answers*.
5. No. Not a method of *computing*. Also not *logically* bound to give the right answer.
6. No. It is not necessary for the existence of an effective method that it should be known to someone.

## 8 Decidable sets

A set is *decidable* if and only if there is an effective method for telling, for each thing that might be a member of the set, whether or not it really is a member.

Some authors require of a formal system that the set of proofs in the system should be decidable. We shall not follow them in this. In each of the main formal systems in this book the set of proofs in the system is in fact decidable. But the metatheory for some of them refers occasionally to systems that may or may not have decidable sets of proofs: it is convenient to call such systems formal systems – and proper too, since they are defined in purely syntactic terms.

Every *finite* set is decidable. Intuitively, think of the members of the set as lined up in a row. Then to determine whether something is a member of the set, check to see whether it is identical with one of the things in the row.

Hereafter we abbreviate 'if and only if' to 'iff'.

EXERCISE

'A set is undecidable iff it has been proved that there is no effective method for telling whether or not a thing is a member of it.' Is there anything wrong with this statement?

ANSWER

Yes. Delete the words 'it has been proved that'.

## 9 1–1 correspondence. Having the same cardinal number as. Having a greater (or smaller) cardinal number than

There is a 1–1 *correspondence* between a set A and a set B iff there is a way (which need not be known to anyone) of pairing off the members of A with the members of B so that

(1) each member of A is paired with exactly one member of B and

(2) each member of B is paired with exactly one member of A. (It follows that no member of either set is left unpaired.)

Two sets are said to *have the same cardinal number* or *the same cardinality* iff there is a 1–1 correspondence between them.

A set A *has a greater cardinal number than* a set B iff there is a 1–1 correspondence between B and a proper subset of A but no 1–1 correspondence between B and the whole of A. (A set C is a *proper subset* of a set D [written: C ⊂ D] iff there is no member of C that is not a member of D but there is a member of D that is not a member of C.)

A set A *has a smaller cardinal number than* a set B iff B has a greater cardinal number than A.

The cardinal number of a set is symbolised by writing two parallel lines over a name of the set. Thus the cardinal number of a set A is $\overline{\overline{A}}$, and '$\overline{\overline{A}} = \overline{\overline{B}}$' means 'The sets A and B have the same cardinal number'. (This notation, which is Cantor's, was meant by him to signify a double abstraction: (1) an abstraction from the nature of the members of the set, (2) an abstraction from the order in which they are taken; i.e. so far as the cardinal number of a set is concerned, neither the nature nor the order of the members of the set matters.)

EXERCISE

'Every set has a 1–1 correspondence with itself.' True or false?

ANSWER

True. The point of including this exercise was to bring out that the word 'pairing' in our definition of 1–1 *correspondence* is used in such a way that a thing can properly be said to be 'paired' with itself.

## 10 Finite sets. Denumerable sets. Countable sets. Uncountable sets

The *natural numbers* are the numbers 0, 1, 2, 3, etc.

A set is *finite* iff it has only a finite number of members; *denumerable* iff there is a 1–1 correspondence between it and the set of natural numbers (so a denumerable set is an infinite set);

*countable* iff it is either finite or denumerable; *uncountable* iff it is neither finite nor denumerable (so an uncountable set is an infinite set).

0 counts as a finite number, so the empty set is a finite set.

The existence of an uncountable set will be proved in §11, assuming the Power Set Axiom (§11.1). (The existence of some other uncountable sets is proved in Appendix 1, without appeal to the Power Set Axiom.)

A denumerable infinity is the smallest sort of infinity: no infinite set has a smaller cardinal number than a denumerable set. So $\aleph_0$ [aleph nought], which is, by definition, the cardinal number of the set of natural numbers, is the smallest transfinite cardinal number. (The cardinal numbers of infinite sets are known as transfinite cardinals.)

It is a characteristic property[1] of any infinite set that there is a 1–1 correspondence between it and at least one of its own proper subsets. For example, there is a 1–1 correspondence between the set of natural numbers and the set of squares of natural numbers, which is a proper subset of the set of natural numbers:

$$\begin{array}{ccccccccc} 0 & 1 & 2 & 3 & 4 & 5 & 6 & 7 & \ldots \\ \updownarrow & \updownarrow & \updownarrow & \updownarrow & \updownarrow & \updownarrow & \updownarrow & \updownarrow & \\ 0 & 1 & 4 & 9 & 16 & 25 & 36 & 49 & \ldots \end{array}$$

This particular correspondence was known to Galileo (1638). The more or less clear recognition that an infinite set can have a 1–1 correspondence with (some of) its own proper subsets seems to go back at least to the Stoics (Chrysippus?) in the third century B.C. For references, see e.g. Kleene (1967, p. 176, fn. 121).

No finite set can have a 1–1 correspondence with any of its own proper subsets. Accordingly C. S. Peirce in 1885 [*Collected Papers*, iii, §402] took the non-existence of such a correspondence as a defining property of finite sets, while Dedekind took its existence as a defining property of infinite ones. (Dedekind published his definition in 1888. He says that he submitted it to

[1] Strictly this holds only under the assumption of the Axiom of Choice, an axiom that commends itself both because of its intuitive plausibility and because it 'has so many important applications in practically all branches of mathematics that not to accept it would seem to be a wilful hobbling of the practicing mathematician' (Mendelson, 1964, p. 201). [For the Axiom of Choice, see §59.]

Cantor in September 1882 and to Schwarz and Weber several years earlier: cf. Dedekind (1887, fn. to §64).)

### EXERCISES

1. Show that the following sets are denumerable:
   (*a*) The set of positive integers, $\{1, 2, 3, 4, \ldots\}$.
   (*b*) The set of even numbers, $\{2, 4, 6, 8, \ldots\}$.
   (*c*) The set of odd numbers, $\{1, 3, 5, 7, \ldots\}$.
   (*d*) The set of integers, $\{\ldots, -3, -2, -1, 0, 1, 2, 3, \ldots\}$.
   (*e*) The set of positive rational numbers.[1]
   (*f*) The set of rational numbers.

2. Show how each of the sets in the last exercise can be put in 1–1 correspondence with one of its own proper subsets.

3. Show that the set of squares of an infinite chessboard of one-inch squares is denumerable.

4. Show that $\aleph_0 - 1 = \aleph_0$ and that $\aleph_0 + 1 = \aleph_0$, and that therefore $\aleph_0 \pm n = \aleph_0$, where $n$ is a natural number.

5. Show that $\aleph_0 . \aleph_0 = \aleph_0$.

### ANSWERS

1. (*a*) Here is the start of a 1–1 correspondence between the set of natural numbers and the set of positive integers:

$$\begin{array}{cccccccc} 0 & 1 & 2 & 3 & 4 & 5 & 6 & \ldots \\ \updownarrow & \updownarrow & \updownarrow & \updownarrow & \updownarrow & \updownarrow & \updownarrow & \\ 1 & 2 & 3 & 4 & 5 & 6 & 7 & \ldots \end{array}$$

In order to prove the existence of a 1–1 correspondence between a set A and the set of natural numbers, it is enough to show how to generate an infinite sequence of members of A that will contain, without repetitions, all the members of A and nothing that is not a member of A. So for the remaining proofs we simply write down the first few terms of appropriate rule-generated sequences, with occasional explanations. [For more on sequences, see §12.]

(*b*) 2, 4, 6, 8, . . .

(*c*) 1, 3, 5, 7, . . .

[1] A *rational number* is a number that can be expressed as a *ratio* of two integers, $\frac{a}{b}$, where $b \neq 0$: e.g. $\frac{1}{2}$, $\frac{39}{17}$, $-15$ $[=\frac{-15}{1}]$.

(*d*) 0, 1, −1, 2, −2, 3, −3, . . . [Here we abandon any attempt to take the members of the set in order of magnitude.]

(*e*) $\frac{1}{1}$, $\frac{1}{2}$, $\frac{2}{1}$, $\frac{1}{3}$, [$\frac{2}{2}$,] $\frac{3}{1}$, $\frac{1}{4}$, $\frac{2}{3}$, $\frac{3}{2}$, $\frac{4}{1}$, $\frac{1}{5}$, [$\frac{2}{4}$,] [$\frac{3}{3}$,] [$\frac{4}{2}$,] $\frac{5}{1}$, $\frac{1}{6}$, . . .

(We write down first every positive rational whose numerator and denominator add up to 2. There is only one, viz. $\frac{1}{1}$. Then we write down every one whose numerator and denominator add up to 3, putting numbers with smaller numerators before numbers with larger ones. Then we do the same for rationals whose numerator and denominator add up to 4, and so on. Each time in this process that we come across a number that has already appeared in the sequence, we omit it: so the numbers in square brackets are not in the sequence; they were put in simply to show how the sequence is obtained.)

(*f*) 0, 1, −1, $\frac{1}{2}$, $-\frac{1}{2}$, 2, −2, $\frac{1}{3}$, $-\frac{1}{3}$, 3, −3, $\frac{1}{4}$, $-\frac{1}{4}$, $\frac{2}{3}$, $-\frac{2}{3}$, $\frac{3}{2}$, $-\frac{3}{2}$, 4, −4, . . .

(The sequence for (*e*) expanded by inserting after each term the corresponding negative number, and adding 0 at the beginning.)

2. In each case simply pair off the first term in the sequence given in the answer to exercise 1 with the second term, the second with the third, and so on. E.g. in case (*a*):

$$\begin{array}{cccccccc} 1 & 2 & 3 & 4 & 5 & 6 & 7 & \ldots \\ \updownarrow & \updownarrow & \updownarrow & \updownarrow & \updownarrow & \updownarrow & \updownarrow & \\ 2 & 3 & 4 & 5 & 6 & 7 & 8 & \ldots \end{array}$$

3. They can be enumerated by starting from an arbitrary square and following the spiral path indicated in the following figure:

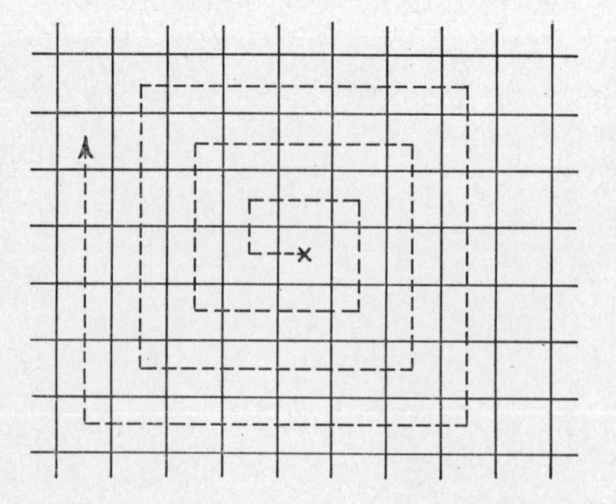

4. Consider the answer to exercise 2.

5. Consider the answer to exercise 3. The number of squares on the infinite chessboard is the product of the number of squares along some arbitrary line on the board and the number of squares along a line at right-angles, which is the product of $\aleph_0$ and $\aleph_0$:

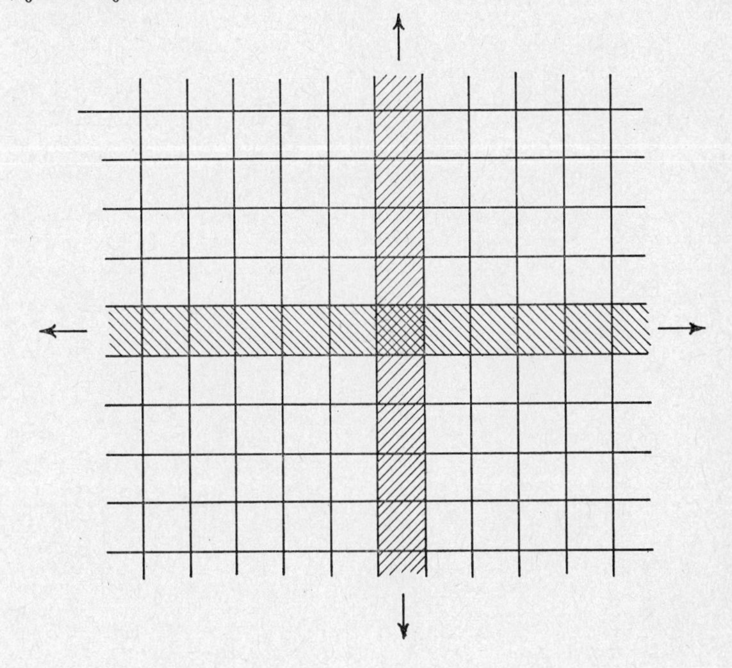

## 11 Proof of the uncountability of the set of all subsets of the set of natural numbers

A set A is a *subset* of a set B [written: A ⊆ B] iff there is no member of A that is not a member of B. The empty set, ∅, is a subset of every set, since for any set C there is no member of ∅ that is not a member of C, simply because there is no member of ∅. Also, every set is a subset of itself. (By contrast, no set is a *proper* subset of itself.)

The set of all subsets of a set A is known as the *power set* of A. It seems intuitively obvious that the set of natural numbers *has* its power set, i.e. that there *is* a set that has for its members all subsets of the set of natural numbers and nothing else. But we have not proved this. It is usual to appeal here to a more general axiom, viz.

11.1 (*The Power Set Axiom*) *For any set there exists its power set*

This axiom cannot be taken to be certainly true, but it looks very plausible, and we shall assume it from now on.

11.2 *The set of all subsets of the set of natural numbers is uncountable*

*Proof.* Clearly the set of all subsets of the set of natural numbers is not finite. For to each natural number there corresponds the set that has that natural number as its sole member; there are denumerably many such sets, and each is a subset of the set of natural numbers.

Now suppose that someone claims that he has found a 1–1 correspondence between the set of natural numbers and the set of all subsets of the set of natural numbers. We shall show how any such claim can be refuted.

Suppose the alleged 1–1 correspondence starts off like this:

| | 0 | 1 | 2 | 3 | 4 | 5 | 6 | 7 | 8 | ... |
|---|---|---|---|---|---|---|---|---|---|---|
| 0 ↔ the set of all natural numbers | Yes | Yes | Yes | Yes | Yes | Yes | Yes | Yes | Yes | ... |
| 1 ↔ the empty set | No | No | No | No | No | No | No | No | No | ... |
| 2 ↔ the set of all even numbers | No | No | Yes | No | Yes | No | Yes | No | Yes | ... |
| 3 ↔ the set of all odd numbers | No | Yes | No | Yes | No | Yes | No | Yes | No | ... |
| 4 ↔ the set of all prime numbers | No | No | Yes | Yes | No | Yes | No | Yes | No | ... |
| 5 ↔ the set of all squares of n.n.s. | Yes | Yes | No | No | Yes | No | No | No | No | ... |
| 6 ↔ the set of all cubes of n.n.s. | Yes | Yes | No | No | No | No | No | No | Yes | ... |
| . . | . | . | . | . | . | . | . | . | . | |
| . . | . | . | . | . | . | . | . | . | . | |
| . . | . | . | . | . | . | . | . | . | . | |

[On the right-hand side of the table we write 'Yes' under a number if it is a member of the set mentioned at its left, and 'No' if it is not.]

We use *Cantor's diagonal argument* to show that our imaginary claimant's supposed 1–1 correspondence is not a 1–1 correspondence after all. For we can define a subset of the set of natural numbers that does not occur in his pairing, viz. the subset defined by starting at the top left-hand corner of the array on the right and going down the diagonal changing each 'Yes' to 'No' and each 'No' to 'Yes', thus:

| 0 | 1 | 2 | 3 | 4 | 5 | 6 | 7 | 8 | ... |
|---|---|---|---|---|---|---|---|---|---|
| **No** | Yes | Yes | Yes | Yes | Yes | Yes | Yes | Yes | ... |
| No | **Yes** | No | No | No | No | No | No | No | ... |
| No | No | **No** | No | Yes | No | Yes | No | Yes | ... |
| No | Yes | No | **No** | No | Yes | No | Yes | No | ... |
| No | No | Yes | Yes | **Yes** | Yes | No | Yes | No | ... |
| Yes | Yes | No | No | Yes | **Yes** | No | No | No | ... |
| Yes | Yes | No | No | No | No | **Yes** | No | Yes | ... |
| . | . | . | . | . | . | . | . | . | |
| . | . | . | . | . | . | . | . | . | |
| . | . | . | . | . | . | . | . | . | |

Going down the diagonal we see that among the members of this subset will be the numbers 1, 4, 5 and 6. The set so defined is a subset of the set of natural numbers that differs from each set in the original pairing in at least one member.

This was only a particular example. It is clear, however, that for *any* alleged 1–1 pairing of the subsets of the set of natural numbers with the natural numbers a similar diagonal argument would yield a subset of the set of natural numbers *not* in the pairing. So we have quite generally:

> There is no 1–1 correspondence between the set of natural numbers and the set of all subsets of the set of natural numbers.

So the set of all subsets of the set of natural numbers is not denumerable. We have seen that it is not finite. So it is uncountable. Q.E.D.

The reader may for a moment think that we could get round the diagonal argument by adding the new subset at the top of the list, pairing it off with the number 0, and shifting each of the other subsets down a place. This would not help. For a fresh application of the diagonal argument to the new list would produce another subset not in the list. And so on, without end. Any attempt to pair off the subsets of the set of natural numbers with the natural numbers leaves out not just one but infinitely many subsets (indeed, uncountably many: cf. 13.6 below).

There is a 1–1 correspondence between the set of natural numbers and the set whose members are {0}, {1}, {2}, {3} and so on. This last set is a proper subset of the set of all subsets of the set of natural numbers. So we have: There is a 1–1 correspondence between the set of natural numbers and a proper subset of the set of all subsets of the set of natural numbers, but no 1–1 correspondence between the set of natural numbers and the set of all subsets of the set of natural numbers. Therefore *the set of all subsets of the set of natural numbers has a greater cardinal number than the set of natural numbers* (cf. the definition of *having a greater cardinal number than* in §9).

By using a general and abstract form of the diagonal argument, Cantor showed that *the power set of a set always has a greater cardinal number than the set itself.* This is known as *Cantor's Theorem.* A proof of it is given in small print below; below that there is an example that may help towards understanding it.

Given the existence of any infinite set whatever, the truth of the Power Set Axiom, and Cantor's Theorem, it follows that there is an unending succession of different and ever greater infinite sets – 'the paradise that Cantor created for us' (Hilbert, 1925).

The following proof of Cantor's Theorem might be skipped at a first reading:

*Cantor's Theorem: The power set of a set has a greater cardinal number than the set itself*

*Proof*

1. Let A be any set. Consider *any* pairing-off of members of A with members of the power set of A that assigns to each distinct member of A a distinct subset of A. Let S be the set of all members of A that are not members of the subset assigned to them. S is a subset of A. But S is not assigned to any member of A. For suppose it were assigned to a member, say $x$, of A. Then $x$ would be a member of S if and only if it were not a member of S. This is a contradiction. So any pairing-off of distinct members of A with distinct members of the power set of A leaves some member of the power set unpaired. So there is no 1–1 correspondence between A and its power set.

2. It remains to show that there is a 1–1 correspondence between A and a proper subset of the power set of A. This is easy. Take as the proper subset the set of all sets that have as their sole member a member of A.

*Example.* Let A be the set {1, 2, 3}. Then the power set of A is the set

$$\{\{1, 2, 3\}, \{1, 2\}, \{1, 3\}, \{2, 3\}, \{1\}, \{2\}, \{3\}, \emptyset\}.$$

A has three members. The power set of A has eight members. There is no 1–1 correspondence between A and its power set; but there is a 1–1 correspondence between A and a proper subset of its power set: take, e.g., the subset {{1}, {2}, {3}}.

Cantor's Theorem is more or less obvious for finite sets. What Cantor did was to show that it held for infinite sets as well.

## 12 Sequences. Enumerations. Effective enumerations

In mathematics a sequence is a function of a certain sort. But we shall use the word in an intuitive way. A *sequence* is an ordering of objects, called the *terms* of the sequence. The same thing may occur more than once in the ordering; e.g. ⟨1, 2, 3, 1⟩ is a sequence of four terms with the number 1 occurring as the first term and as the fourth term. A sequence $s$ is the same as a sequence $s'$ iff $s$ and $s'$ have exactly the same number of terms and the first term of $s$ is the same as the first term of $s'$, the second term of $s$ is exactly the same as the second term of $s'$, and so on. So, e.g., $\langle 1, 2, 1\rangle \neq \langle 1, 2\rangle$, and $\langle 1, 2, 3\rangle \neq \langle 3, 2, 1\rangle$.

A sequence of $n$ terms is also known as an $n$-tuple.

A sequence may be finite or infinite. A denumerable sequence is a sequence with denumerably many terms, and may be symbolised by giving the first few terms and then putting dots; e.g. ⟨1, 2, 3, . . .⟩ and ⟨1, 1, 1, . . .⟩ and ⟨1, 2, 1, 2, 1, 2, . . .⟩ and ⟨4, 9, 13, 5, 5, 5, . . ., 5, . . .⟩ are denumerable sequences. In the last one every term from the fourth on is the number 5.

In the examples given above all the terms in all the sequences are numbers. But just as there are sets with things other than numbers as members, so there are sequences with things other than numbers as terms.

An *enumeration* of a set A is a finite or denumerable sequence of which every member of A is a term and every term is a member of A. For example, the sequences ⟨3, 1, 2⟩ and ⟨1, 2, 3, 1⟩ are both enumerations of the set {1, 2, 3}.

An *effective enumeration* is an enumeration which is finite or for which there is an effective method for telling what the *n*th term is, for each positive integer *n*. (Every finite enumeration is effective, because there is an effective method – remember it need not be known to anyone – for enumerating the members of any finite set.)

EXERCISES

1. 'If there is an enumeration with repetitions of a set A, then there is an enumeration without repetitions of the set A.' True or false?
2. (*a*) If in exercise 1 above 'effective enumeration' was substituted for 'enumeration' both times, would the resulting statement be true?
   (*b*) If it would be, describe a method for calculating the *n*th term of the new enumeration. If not, say why not.

ANSWERS

1. True. Simply delete repetitions.
2. (*a*) Yes.
   (*b*) Calculate the first term of the old sequence. Put it down as the first term of the new. Calculate the second term of the old. Put it down as the second term of the new *unless* it is identical with an earlier term in the new sequence. After each addition to the new sequence, check how many terms you have put in it. Sooner or later you are bound to come to the *n*th term of the new sequence (if there is an *n*th term), and you will know it to be the *n*th term.

## 13 Some theorems about infinite sets

13.1 *Any subset of a countable set is countable*

*Proof.* Let A be an arbitrary countable set and B be an arbitrary subset of A.

(*a*) If A is finite, then obviously B is finite and so countable.

(*b*) If A is denumerable, then (by definition) there is a 1–1 correspondence between it and the set of natural numbers, and so there is a denumerable sequence that enumerates A without repetitions. Delete from this sequence all terms that are not members of B, and the result is a finite or denumerable sequence that enumerates B without repetitions. So B is countable.

The *union* of two sets A and B [written: A ∪ B] is the set that has for its members all the members of A and all the members of B.

13.2 *The union of a denumerable set and a finite set is denumerable*

*Proof.* Let A be any denumerable set and B be any finite set. Let B have exactly $n$ members. Then the members of B can be paired off with the first $n$ natural numbers (viz. 0, 1, . . ., $n - 1$), and the members of A with the natural numbers from $n$ on, first deleting any members of A that are also members of B.

13.3 *The union of a denumerable set and a denumerable set is a denumerable set*

*Proof.* Take members from each set alternately, deleting any repetitions. Example: Let A be the set of prime numbers, {2, 3, 5, 7, 11, 13, 17, . . .}, and B be the set of odd numbers, {1, 3, 5, 7, 9, 11, 13, 15, . . .}. Then the union of A and B can be enumerated in the following way:

⟨2, 1, 3, ~~3~~, 5, ~~5~~, 7, ~~7~~, 11, 9, 13, ~~11~~, 17, ~~13~~, 19, 15, . . .⟩

Given the proofs of 13.2 and 13.3, the proofs of the next two theorems are easy, and they are left to the reader:

13.4 *The union of a countable set and a finite set is countable*

13.5 *The union of a countable set and a countable set is countable*

*A – B* is the set that has for its members all those members of A that are not members of B.

13.6 *The removal from an uncountable set of countably many members leaves an uncountable set remaining*

*Proof*

(*a*) [The case where finitely many members are subtracted.]

Let A be any uncountable set, and B be any finite set of members of A. Suppose A – B were countable. Then by 13.4 the union of A – B with B would be countable. But the union of A – B with B is A itself, and by hypothesis A is uncountable. So A – B must be uncountable.

(*b*) [The case where denumerably many members are subtracted.] Let A be any uncountable set, and B be any denumerable set of members of A. Suppose A – B were countable. Then by 13.5 the union of A – B with B would be countable. But the union of A – B with B is A itself, and by hypothesis A is uncountable. So A – B must be uncountable.

## 14 Informal proof of the incompleteness of any finitary formal system of the full theory of the natural numbers

By 'finitary formal system' we shall mean a formal system with a finite or denumerable alphabet of symbols, wffs only finitely long, and rules of inference (if any) using only finitely many premisses. (In recent years logicians have worked on systems with uncountable alphabets, with infinitely long wffs, with rules of inference having infinitely many premisses, etc.)

A formal system of the full theory of the natural numbers is a formal system (some or all of) whose theorems can be interpreted as expressing truths of the full theory of the natural numbers. Such a system will be said to be *incomplete* if there are truths of the full theory of the natural numbers that are not theorems of the system [i.e. truths that are not expressed by any theorem of the system, on its intended interpretation].

The full theory of the natural numbers is here taken to include, among other things, all truths to the effect that a particular natural number is a member of some particular set of natural numbers. Remember that a set A is identical with a set B if and only if A has exactly the same members as B, and that this identity is not affected by quite radical differences in the specifications of A and B. [E.g. if the only number I am now thinking of is the number 17, then the set of numbers-I-am-now-thinking-of is identical with the set whose sole member is the number 17.]

14.1 *Any finitary formal system has only countably many wffs and therefore only countably many theorems*

*Proof*

Stage 1. A denumerable alphabet has no greater powers of expression than a finite alphabet, or even than a two-symbol alphabet. For suppose we have a denumerable alphabet with symbols $a_1, a_2, a_3, \ldots$ Then there is a 1–1 correspondence between it and the set $\{10, 100, 1000, \ldots\}$ of strings of symbols from the alphabet whose only symbols are 0 and 1; and these strings from the finite alphabet can be used to do whatever can be done with the symbols from the denumerable alphabet.

Stage 2. The set of distinct finitely long wffs that can be got from a finite alphabet is countable. *Proof.* Replace each symbol of the alphabet by a '1' followed by a string of one or more zeros, as in Stage 1. Each wff then becomes a numeral composed exclusively of ones and zeros and beginning with a one.[1] To each distinct wff there corresponds a distinct numeral. Each of these distinct numerals stands for a distinct natural number. There are only denumerably many natural numbers, so there are at most denumerably many wffs. So the set of wffs of any finitary formal system is countable. In any formal system every theorem is a wff. So the set of theorems of any finitary formal system is also countable.

14.2 *There are uncountably many truths of the full theory of the natural numbers*

*Proof.* To each of the uncountably many subsets of the set of natural numbers [see 11.2] there corresponds a distinct truth of the full theory of the natural numbers, viz. the truth that zero[2] is (or is not, as the case may be) a member of that subset. Therefore there are at least as many truths of the full theory of the

[1] E.g. suppose that the alphabet consists of the four symbols

$$p \quad \supset \quad ( \quad )$$

Replace these in any wff by

$$10 \quad 100 \quad 1000 \quad 10000$$

respectively. Then (e.g.) the wff

$$(p \supset p)$$

becomes

$$1000101001010000.$$

[2] There is nothing special about zero here; any other natural number would do.

natural numbers as there are subsets of the set of natural numbers, and so there are uncountably many such truths.

14.3 *Any finitary formal system of the full theory of the natural numbers is incomplete, in the sense explained* [*see paragraph 2 of this section*]

*Proof.* By 14.1 any finitary formal system of the full theory of the natural numbers has only countably many (formal) theorems. On the intended interpretation each formal theorem will have just one definite unambiguous meaning and, if it is true on the intended interpretation, it will express just one truth. For simplicity's sake we shall assume that on the intended interpretation each distinct theorem expresses a distinct truth of the full theory of the natural numbers. [There are other possibilities: e.g. two distinct theorems might express the same truth, or some theorems might express falsehoods, or some might express truths that were not truths of the full theory of the natural numbers. A full proof would cover these possibilities, using the theorem 13.1 that any subset of a countable set is countable.] Also for simplicity's sake we shall identify interpreted theorems with the truths they express. Then we get: Any finitary formal system of the full theory of the natural numbers will include among its interpreted theorems only countably many truths of the full theory of the natural numbers. But by 14.2 there are uncountably many such truths. So by 13.6 any finitary formal system will fail to include among its interpreted theorems uncountably many such truths. Therefore any finitary formal system of the full theory of the natural numbers is incomplete.

Q.E.D.

## Appendix 1: Intuitive theory of infinite sets and transfinite cardinal numbers

Proofs informal throughout. Some theorems left unproved. The material is not essential to the rest of the book.

The set of *real numbers* has as members all the rational numbers [explained in §10, exercise 1, fn.] and all the irrational numbers, but excludes complex numbers (originally called 'imaginary numbers': hence the word 'real' in 'real numbers'). Complex numbers are numbers involving $i$, the square root of $-1$: e.g. $3+2i$. The irrational numbers are numbers (other than complex or transfinite numbers) that cannot be expressed as a ratio of two integers: examples are $\sqrt{2}$ and $\pi$.

The (real) *continuum* is the set of real numbers.

The *linear continuum* is the set of all points on a line. For the moment take 'line' here to mean 'infinite line'. We shall see later (Theorem A8) that in fact it does not matter how long the line is.

There is a well-known 1–1 correspondence between the set of real numbers and the set of all points on an infinite line:

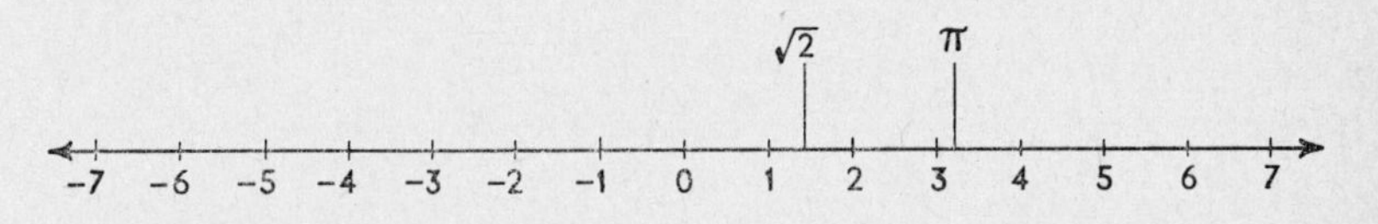


So we have:

*Theorem* A1. *The cardinal number of the* (*real*) *continuum is the same as the cardinal number of the linear continuum*

So from now on we may speak simply of 'the cardinal number of the continuum'.

*Notation. The cardinal number of the continuum is* $\mathfrak{c}$ [the lower-case German 'c']

*Theorem* A2. *The set of real numbers is uncountable* [Cantor, 1873]

*Proof.* Three stages:

1. Every real number can be uniquely represented by a non-terminating decimal numeral. Those that it would be natural to represent by terminating decimals are to be represented by non-terminating equivalents: e.g. the real number $3\frac{1}{4}$ by 3·24999999 . . ., not by 3·25; the number 1 by 0·99999 . . . .

2. We concentrate first on the set of real numbers greater than 0 and less than or equal to 1. This set is clearly not finite. Now suppose that each distinct natural number is paired off with a distinct real number, represented by a non-terminating decimal. It is easy to show by a diagonal argument that any

such pairing leaves some real number unpaired. For any such pairing will begin like this:

$$\begin{array}{ll} 0 & 0 \cdot d\,d\,d\,d\,d\,d\,d \ldots \\ 1 & 0 \cdot d\,d\,d\,d\,d\,d\,d \ldots \\ 2 & 0 \cdot d\,d\,d\,d\,d\,d\,d \ldots \end{array}$$

where each *d* is some digit or other (i.e. numeral from 0 to 9), possibly all the same, possibly not. And it is clear that any numeral got by starting off with a zero, then a decimal point, then a digit different from the first *d* on the diagonal (but not 0), then a digit different from the second *d* on the diagonal (but not 0), and so on, will represent a real number greater than 0 and less than or equal to 1 that is not paired off with any natural number. So the set of real numbers greater than 0 and less than or equal to 1 is uncountable.

3. The set of real numbers greater than 0 but less than or equal to 1 is a subset of the set of real numbers. We have just shown that it is uncountable. By 13.1 any subset of a countable set is countable. So the set of (all) real numbers is uncountable.

Q.E.D.

Since the set of natural numbers is a proper subset of the set of real numbers, we have as a corollary of Theorem A2:

*Theorem* A3. $\mathfrak{c} > \aleph_0$

We shall make a good deal of use of the following two theorems, which we state without proof:

*Theorem* A4. *The addition to, or subtraction from, an infinite set of finitely many things yields a set with the same cardinal number as the original*

This is an extension of 13.2, which covered only denumerable sets. A4 covers uncountable sets as well, and its proof involves the Axiom of Choice, which is beyond the scope of this book. [For the Axiom of Choice, see §59.]

*Theorem* A5. *The subtraction of a denumerable set from an uncountable set leaves a set with the same cardinal number as the uncountable set*

Cf. 13.6. A5 gives us more definite information about the cardinal number of the resulting set than 13.6.

*Theorem A6. Each of the following sets has the cardinal number of the continuum:*

*A. The set of real numbers x such that* $0 \leqslant x \leqslant 1$
*B. The set of real numbers x such that* $0 < x \leqslant 1$
*C. The set of real numbers x such that* $0 \leqslant x < 1$
*D. The set of real numbers x such that* $0 < x < 1$
*E. The set of non-negative real numbers*

*Proof*

(Sets *A*, *B*, *C*, *D*) *A* is *B* with one extra thing, viz. the number 0. *C* is *A* with one member less, viz. the number 1. *D* is *C* with one member less, viz. the number 0. We have already proved that *B* has the cardinal number of the continuum [in the course of proving Theorem A2]. So *A* has, by Theorem A4. So *C* has, by Theorem A4 applied to the result for *A*. So *D* has, by A4 applied to the result for *C*.

(Set *E*) *E* has *D* as a proper subset. So if there is no 1–1 correspondence between *E* and *D*, $\overline{\overline{D}} < \overline{\overline{E}}$ (by the definition of < for transfinite cardinals); and if there *is* a 1–1 correspondence, $\overline{\overline{D}} = \overline{\overline{E}}$. But either there is or there is not a 1–1 correspondence between *E* and *D*. So $\overline{\overline{D}} \leqslant \overline{\overline{E}}$. Similarly $\overline{\overline{E}} \leqslant \overline{\overline{R}}$, where *R* is the set of all real numbers. But $\overline{\overline{D}} = \overline{\overline{R}} = \mathfrak{c}$. So we have:

$$\mathfrak{c} = \overline{\overline{D}} \leqslant \overline{\overline{E}} \leqslant \overline{\overline{R}} = \mathfrak{c}.$$

So $\mathfrak{c} \leqslant \overline{\overline{E}} \leqslant \mathfrak{c}$.

So $\overline{\overline{E}} = \mathfrak{c}$.

(A geometrical proof of the same result is given below: Theorem A8*H*.)

*Theorem* A7. *The set of all subsets of the set of natural numbers* (*the power set of the set of natural numbers*) *has the cardinal number of the continuum*

*Proof*

Every set of natural numbers can be uniquely represented by a denumerable string of 'Yes'es and 'No's, as in the tables in §11. For example, the string that begins with

Yes Yes No No Yes No

and goes on with nothing but 'No's represents the set {0, 1, 4}. So there is a 1–1 correspondence between the set of all subsets of the set of natural numbers and the set of denumerable strings of 'Yes'es and 'No's. And there is a 1–1 correspondence between this last set and the set of denumerable strings of '1's and '0's. For example, the string of 'Yes'es and 'No's mentioned above corresponds to the string that begins with

$$1\ 1\ 0\ 0\ 1\ 0$$

and goes on with nothing but '0's.

Put the binary equivalent of a decimal point in front of a denumerable string of '1's and '0's and you get an expression that, in the binary system, denotes a real number $\geqslant 0$ and $\leqslant 1$. For example, the expression that starts off with

$$\cdot 1\ 1\ 0\ 0\ 1\ 0$$

and goes on with nothing but '0's denotes the real number

$$\tfrac{1}{2} + \tfrac{1}{4} + \tfrac{0}{8} + \tfrac{0}{16} + \tfrac{1}{32} + \tfrac{0}{64} + 0$$

i.e. the real number $\frac{25}{32}$. Each denumerable string of '1's and '0's, preceded by a binary point, represents a real number $\geqslant 0$ and $\leqslant 1$, and each real number $\geqslant 0$ and $\leqslant 1$ is represented by some string. But it is not a 1–1 correspondence, since some numbers are represented by more than one string. For example, the number $\frac{1}{4}$ is represented both by

$$\cdot 0\ 1\ 0\ 0\ 0\ 0\ 0 \ldots$$

and by

$$\cdot 0\ 0\ 1\ 1\ 1\ 1\ 1 \ldots$$

However, if we took away from the set of strings the set of all strings that from some point on consist wholly of '0's, there *would* be a 1–1 correspondence between the set of strings that was left and the set of real numbers $>0$ and $\leqslant 1$. But the set of all denumerable strings of '1's and '0's that from some point on consist wholly of '0's can be *enumerated* (we leave this to the reader as an exercise). And the subtraction of a denumerable set from an uncountable set leaves a set with the same cardinality as the uncountable set [Theorem A5]. So we have, using '$\simeq$' for 'has a 1–1 correspondence with':

> The set of subsets of the set of natural numbers $\simeq$ The set of denumerable strings of 'Yes'es and 'No's $\simeq$ The set of denumerable strings of '1's and '0's $\simeq$ The set of denumerable

strings of '1's and '0's other than those strings that from some point on consist wholly of '0's $\simeq$ The set of real numbers $>0$ and $\leqslant 1 \simeq$ The set of real numbers.

So the set of subsets of the set of natural numbers has the cardinal number of the continuum.

Q.E.D.

*Theorem* A8. *Each of the following sets has the cardinal number of the continuum*:

*F. The set of points on a line one arbitrary unit long*
*G. The set of points on a line two arbitrary units long*
*H. The set of points on the (infinite) half-line*
*I. The set of points on the (infinite) line*

*Proof*

*F.* Use Theorem A6*A* and the familiar 1–1 correspondence illustrated by the figure (let *AB* be one arbitrary unit long):

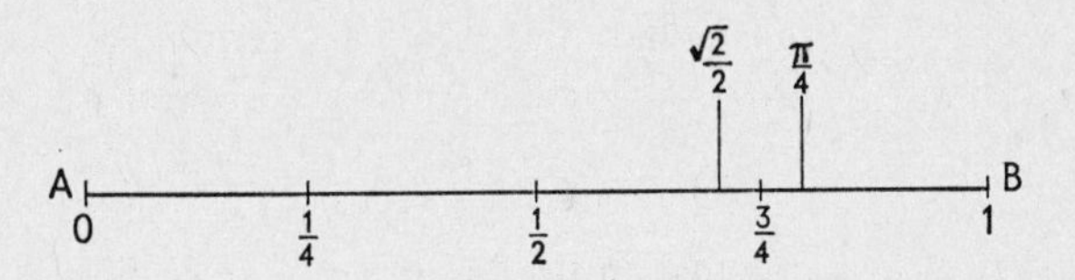


*G.* One proof is by letting the line *AB* in the proof for *F* be two units long. More intuitive is the following geometrical proof:

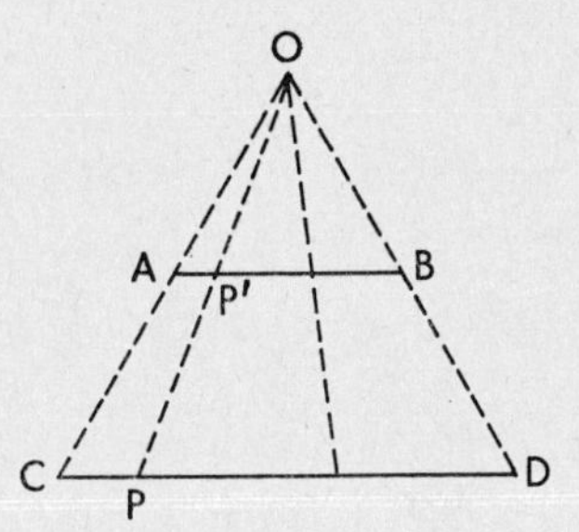


To each point on the two-unit-long line *CD* there corresponds a unique point on the one-unit-long line *AB*, and vice versa. For example, to the point *P* on *CD* there corresponds the unique point *P′* on *AB*, and vice versa.

*H.* One proof is by the familiar 1–1 correspondence between the set of points on the (infinite) half-line and the set of non-negative real numbers (already shown to have the cardinal c: Theorem A6*E*):

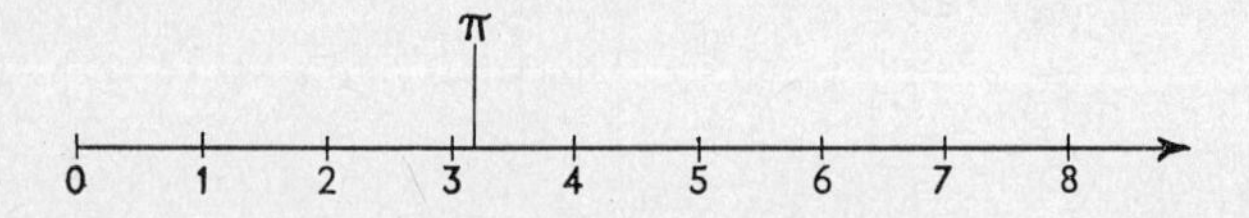


Another is geometrical:

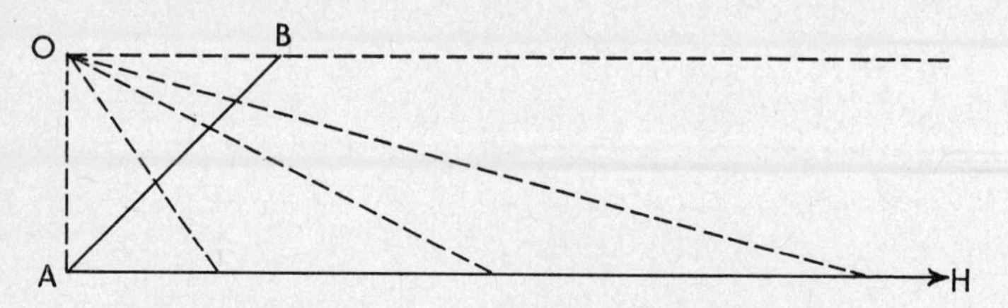


There is a 1–1 correspondence between the points on the infinite half-line *AH* and the points on *AB* other than *B* (no point on *AH* corresponds to the point *B*, for *OB* is parallel to *AH*). By Theorem A4 the number of points on *AB* is the same as the number of points on *AB* other than *B*. So there is the same number of points on the infinite half-line as on a finite line.

*I*. Proved already (Theorem A1). It could also be proved by the following two figures, with the help of Theorem A8*F*, e.g., [let *XY* be one unit long] and A4 [the number of points on the angled line *CAB* is the same as the number of points on *CAB* other than the points *B* and *C*]:

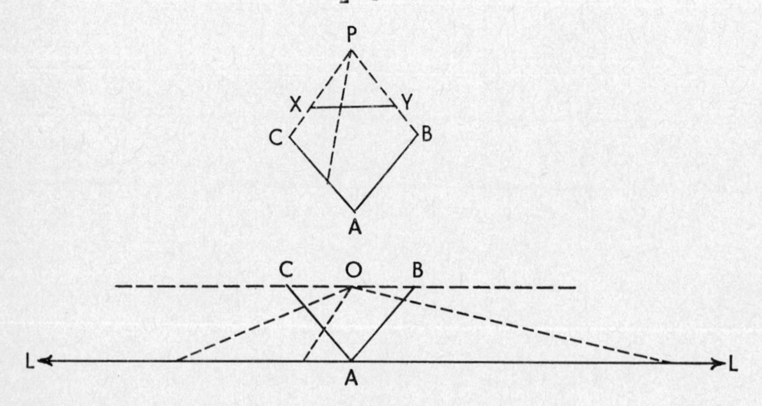

There is the same number of points on the one-unit-long line *XY* as there is on the angled line *CAB*, and the same number on *CAB* as on *L* [using Theorem A4].

*Theorem* A9. *Each of the following sets has the cardinal number of the continuum*:

*J. The set of all points of a square* [*Cantor*, 1877]
*K. The set of all points in a cube*
*L. The set of all points in an infinite plane*
*M. The set of all points in infinite three-dimensional Euclidean space*

*Proof*

*J*. Four stages:

1. Let *ABCD* be a square with sides one arbitrary unit long:

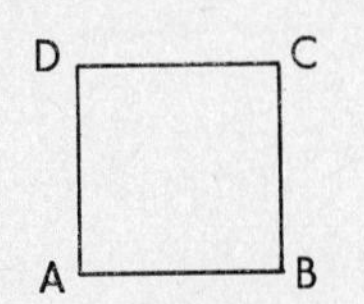


We concentrate on the points of the square other than those on the sides *AB* or *AD*. If we can show that the set of all remaining points of the square has the cardinal number $\mathfrak{c}$, then the union of that set with the set of all points on *AB* will also have the cardinal number $\mathfrak{c}$, and the union of this last set with the set of all points on *AD* will also have the cardinal number $\mathfrak{c}$. For the union of two sets each having the cardinal number of the continuum has the cardinal number of the continuum. [As an illustration, consider a line two units long:

The set of points on *XY* has the cardinal number $\mathfrak{c}$. The set of points on *YZ* has the cardinal number $\mathfrak{c}$. The union of these sets is the set of all points on *XZ*, which also has the cardinal number $\mathfrak{c}$. So $\mathfrak{c} + \mathfrak{c} = \mathfrak{c}$.]

In 2, 3 and most of 4 below we shall write, for brevity, 'point of the square' meaning 'point of the square not on *AB* or *AD*'.

2. The first basic idea of the proof is that any point of the square *ABCD* corresponds to a pair of non-terminating decimals

denoting real numbers $>0$ and $\leqslant 1$, viz. the numbers that are the Cartesian co-ordinates of the point when for axes we choose *AB* and *AD*:

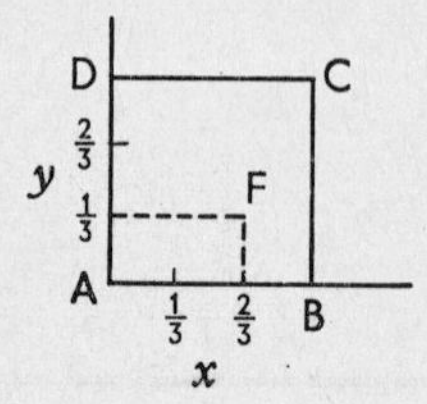


In our illustration the co-ordinates of *F* are $x = 0{\cdot}6666\ldots$, $y = 0{\cdot}3333\ldots$

3. The second basic idea is to reduce these two non-terminating decimals to a single non-terminating decimal by interlacing their digits. In our illustration the new decimal will be 0·6363636363 . . . So to each distinct point of the square there corresponds a distinct non-terminating decimal denoting a real number $>0$ and $\leqslant 1$.

4. But there are some non-terminating decimals which do not split up into two non-terminating decimals, viz. all those decimals in which the digit 0 occurs alternately and infinitely many times from some point on: e.g. 0·6360606060606060 . . ., which splits up into 0·6666 . . . and 0·3. Terminating decimals were not allowed in the correspondence we set up in Stage 2. So we have not yet got a 1–1 correspondence between the points of the square and the non-terminating decimals denoting real numbers $>0$ and $\leqslant 1$: to each distinct point of the square there corresponds a distinct non-terminating decimal denoting a real number $>0$ and $\leqslant 1$, but not vice versa. To deal with this complication we need the third basic idea of the proof. In getting back from the single decimal to the pair of decimals we take single digits alternately as before *unless* the digit is a zero: if it is a zero, we take the *group* of digits that begins with the zero and ends with the first digit that is not a zero. So

$$0{\cdot}6|3|6|06|06|06|06|\ldots$$

splits up into

$$0{\cdot}660606\ldots$$

and

$$0{\cdot}30606\ldots$$

So terminating decimals are avoided and we get our 1–1 correspondence between the set of points of the square (other than the points on *AB* or *AD*) and the set of all non-terminating decimals denoting real numbers $>0$ and $\leqslant 1$. We know already that this last set has the cardinal number $\mathfrak{c}$. In view of the remarks in Stage 1, it follows that the set of *all* points of the square has the cardinal number $\mathfrak{c}$.

Q.E.D.

(For remarks on the trouble Cantor had in getting this result, see Fraenkel (1961, p. 103).)

*K*. As for *J*, but interlacing *three* non-terminating decimals, the decimals denoting the co-ordinates of the point in the (three-dimensional) cube.

*L*. Think of the plane as the infinite chessboard of §10, exercise 3, which we there showed to have only denumerably many one-inch squares. We have already seen that the union of two sets each of which has the cardinal number $\mathfrak{c}$ has the cardinal number $\mathfrak{c}$ [Proof of Theorem A9*J*, Stage 1]. So the union of $n$ sets each of which has the cardinal number $\mathfrak{c}$ will also have the cardinal number $\mathfrak{c}$. But the same is true for the union of denumerably many sets each of which has the cardinal number $\mathfrak{c}$. [As an illustration, consider the infinite line divided into unit segments:

There are denumerably many segments each having $\mathfrak{c}$ points. So the line has $\aleph_0 \, . \, \mathfrak{c}$ points. So $\aleph_0 \, . \, \mathfrak{c} = \mathfrak{c}$.] The plane consists of the denumerably many one-inch squares, and the number of points in it is therefore $\aleph_0 \, . \, \mathfrak{c}$, which is $\mathfrak{c}$. So there is the same number of points in a plane as there is on (e.g.) a one-inch line.

*M*. As for *L*, but this time with one-inch cubes, and instead of a simple two-dimensional spiral path we take a more complicated three-dimensional path.

Later proofs can be shortened by using the following theorems of the arithmetic of transfinite cardinals:

*Theorem* A10. $\aleph_0 . \aleph_0 = \aleph_0$

*Proof.* Cf. §10, answer to exercise 5.

*Theorem* A11. $c = c . c = c^2 = c . c . c = c^3 = . . . = c^n$, *where n is any positive integer*

*Proof.* For $c^2 = c$ consider the result A9*J*. [The number of the set of all points in a square is the product of the number of points on one side and the number of points on an adjacent side $= c . c = c$.] For $c^3 = c$ consider A9*K*. And so on.

*Theorem* A12. *If a set has the finite cardinal number n, then its power set has the finite cardinal number* $2^n$

Illustration:

| | | Members of the set | | |
|---|---|---|---|---|
| | | 1 | 2 | 3 |
| Subsets of the set | 1. {1, 2, 3} | Yes | Yes | Yes |
| | 2. {2, 3} | No | Yes | Yes |
| | 3. {1, 3} | Yes | No | Yes |
| | 4. {3} | No | No | Yes |
| | 5. {1, 2} | Yes | Yes | No |
| | 6. {2} | No | Yes | No |
| | 7. {1} | Yes | No | No |
| | 8. The empty set, ∅ | No | No | No |

$2^3 = 8$.

*Theorem* A13. (Generalisation of A12 to transfinite cardinals.)[1] *If a set has the transfinite cardinal number* $\alpha$, *then its power set has the transfinite cardinal number* $2^\alpha$

*Theorem* A14. $2^\alpha > \alpha$, *for each transfinite cardinal number* $\alpha$

*Proof.* From Cantor's Theorem (proved in §11) and Theorem A13.

*Theorem* A15. *The power set of the set of natural numbers has the cardinal number* $2^{\aleph_0}$

*Proof.* From A13 and the definition of $\aleph_0$.

*Theorem* A16. $c = 2^{\aleph_0}$

*Proof.* From A7 and A15.

[1] The justification of Theorem A13 is more complicated than this remark suggests: see, e.g., Fraenkel (1961, chap. II, §7).

*Theorem* A17. *The set of all points in $\aleph_0$-dimensional space has the cardinal number of the continuum*

*Proof.* The set of all points in three-dimensional space (e.g.) has the number $\mathfrak{c}^3$. The set of all points in $\aleph_0$-dimensional space has the number $\mathfrak{c}^{\aleph_0}$. Then 'with a few strokes of the pen' (Cantor) we get:

$$\mathfrak{c}^{\aleph_0} = (2^{\aleph_0})^{\aleph_0} \text{ [A16] } = 2^{\aleph_0 \cdot \aleph_0} = 2^{\aleph_0} \text{ [A10] } = \mathfrak{c}.$$

Or in other words: The number of points in infinite space of $\aleph_0$ dimensions is the same as the number of points on a line a billionth of an inch long.

Cantor conjectured that there is no cardinal number $\alpha$ such that $\aleph_0 < \alpha < \mathfrak{c}$. This conjecture is known as

*The Continuum Hypothesis. There is no cardinal number greater than $\aleph_0$ and smaller than $\mathfrak{c}$ [$2^{\aleph_0}$]*

In 1938 Kurt Gödel showed that the Continuum Hypothesis cannot be disproved from the usual axioms of set theory, and in 1963 Paul Cohen showed that it cannot be proved from them either. The Continuum Hypothesis is thus *independent* of the usual axioms of set theory. (These results are under the hypothesis that the usual axioms of set theory are consistent. So far nobody has proved that they are, but most workers in the field think that they are.)

*The Generalised Continuum Hypothesis* (which implies the Continuum Hypothesis). *For each transfinite cardinal number $\alpha$, there is no cardinal number greater than $\alpha$ and smaller than $2^\alpha$*

EXERCISE

Are there sets with cardinal numbers greater than $\mathfrak{c}$?

ANSWER

Yes, assuming the Power Set Axiom [i.e. the axiom that for each set there exists its power set]. E.g. the set of all subsets of the set of real numbers has the cardinal number $2^{\mathfrak{c}}$, which is greater than $\mathfrak{c}$. The power set of the power set of the set of natural numbers also has the cardinal number $2^{\mathfrak{c}}$. The power sets of these sets have the still greater cardinal number, $2^{(2^{\mathfrak{c}})}$, and so on.

# PART TWO

# Truth-functional Propositional Logic

*Truth-functional propositional logic* is the theory of (1) those logical truths that can be expressed using only unanalysed propositional variables and truth-functional propositional connectives, and (2) principles of inference relating things that can be expressed by those means.

A *proposition* (as the word is used in this book; others define it differently) is a *sentence* expressing something true or false. It is an abstract thing; its tokens are strings of words.

## 15 Functions

A function is a *relation* that satisfies certain conditions. (A function is an abstract thing. It is not to be identified with any linguistic expression. There are, for example, uncountably many functions and only countably many actual or possible linguistic expressions.)

For simplicity, we confine ourselves for the moment to *functions of one argument* and *two-termed relations.*

The *domain* of a two-termed relation is the set of all things that have the relation to something or other. The *range* of a two-termed relation is the set of all things to which something or other has the relation.

Example 1. The domain of the relation of *being husband of* is the set of all husbands; the range is the set of all wives.

*Definition.* A *function of one argument* is a two-termed relation that assigns to each member of its domain *one and only one* member of its range.

Examples:

2. The two-termed relation *being husband of* is not a function, for in some countries a husband may have more than one wife. By contrast the relation *being monogamously married to* is a function.

3. The relation $f$ that has for its domain the set of positive integers and for its range the set of even numbers and is defined by the rule

$$f(x) = 2x$$

is a function. (Each positive integer is assigned one and only one number that is the product of the given integer and 2.)

4. The relation of *having as father* (where father = human biological father) is a function. Each thing that has a father has one and only one father. Two or more things can have the same father, but this does not prevent the relation from being a function. What matters is that no member of the domain is assigned more than one member of the range.

We give examples to show what is meant by '*a function of n arguments*'.

5. The function $g$ defined by the rule

$$g(x, y) = x + y$$

is a function of *two* arguments.

6. The function $h$ defined by the rule

$$h(x, y, z) = (x \cdot y) + z$$

is a function of *three* arguments.

7. The function $j$ defined by the rule

$$j(x, y) = (x^2 + y)^x$$

is a function of *two* arguments.

So a *function of n arguments* ($n > 1$) is a function whose domain is a set of *n-termed sequences* or *n-tuples*. By identifying a thing with the sequence of which it is the sole term, we can speak also, as we have been doing, of a function of *one* argument.

8. The function $k$ defined by the rule

$$k(x) = x + 1$$

is a function of *one* argument.

*'Arguments' and 'values' of a function*

'Arguments' and 'values' are used as follows:
Consider the function $m$ defined by the rule

$$m(x, y) = x \cdot y$$

and having for its domain the set of ordered pairs (two-termed sequences) of natural numbers and for its range the set of natural numbers. For the arguments $x = 3$, $y = 4$ this function has the value 12. For the arguments $x = 15$, $y = 0$ this function has the value 0. So the set of *values* of a function is simply the range of the function. An *argument* of a function is a term of a sequence that belongs to the domain of the function. So the set of arguments of a function does not coincide with the domain of the function except in the case of functions of one argument.

A function whose arguments and values are natural numbers is said to be a function from natural numbers to natural numbers.

*Definition. f is the same function as g*

Just as a set A is the same set as a set B if and only if it has exactly the same members as B, no matter how the members are specified or described, so *a function f is the same function as a function g* if and only if

(1) $f$ and $g$ have the same domain

and

(2) $f$ and $g$ have the same value for the same $n$-tuple of arguments, for each $n$-tuple in the domain.

What this amounts to is that a function $f$ and a function $g$ can be the same function even if described in wildly different terms. All that matters for the identity of $f$ and $g$ is (1) the identity of their domains, and (2) the identity of the things picked out by $f$ and $g$ for each member of the domain.

## 16 Truth functions

We call truth and falsity *truth values* (and in this book we allow nothing else to be a truth value). 'T' denotes the truth value truth; 'F' denotes the truth value falsity.

*Definition.* A *truth function* is a *function whose domain is a set of sequences of truth values and whose range is a subset of the set of truth values, i.e. the set {T, F}*. Or in other words:
*A truth function is a function whose arguments and values are truth values.*

Examples:

1. The function $q$ that has as its domain the set of all two-termed sequences whose terms are from the set {T, F} and that has as its range the set {T, F}, and that is defined by the rule

$$\begin{cases} q(\mathrm{T}, \mathrm{T}) = \mathrm{T} \\ q(\mathrm{F}, \mathrm{T}) = \mathrm{F} \\ q(\mathrm{T}, \mathrm{F}) = \mathrm{F} \\ q(\mathrm{F}, \mathrm{F}) = \mathrm{F} \end{cases}$$

is a truth function, viz. *conjunction*.

2. The function $r$ that has truth values as arguments and values and is defined by the rule

$$\begin{cases} r(\mathrm{T}, \mathrm{T}) = \mathrm{T} \\ r(\mathrm{F}, \mathrm{T}) = \mathrm{T} \\ r(\mathrm{T}, \mathrm{F}) = \mathrm{F} \\ r(\mathrm{F}, \mathrm{F}) = \mathrm{T} \end{cases}$$

is a truth function, viz. *material implication.*

3. The function $s$ that has truth values as arguments and values and is defined by the rule

$$\begin{cases} s(\mathrm{T}) = \mathrm{F} \\ s(\mathrm{F}) = \mathrm{T} \end{cases}$$

is a truth function, viz. *negation.*

4. The function $t$ that has truth values as its arguments and values and is defined by the rule

$$\begin{cases} t(\mathrm{T}, \mathrm{T}, \mathrm{T}) = \mathrm{T} \\ t(\mathrm{F}, \mathrm{T}, \mathrm{T}) = \mathrm{F} \\ t(\mathrm{T}, \mathrm{F}, \mathrm{T}) = \mathrm{F} \\ t(\mathrm{F}, \mathrm{F}, \mathrm{T}) = \mathrm{F} \\ t(\mathrm{T}, \mathrm{T}, \mathrm{F}) = \mathrm{F} \\ t(\mathrm{F}, \mathrm{T}, \mathrm{F}) = \mathrm{F} \\ t(\mathrm{T}, \mathrm{F}, \mathrm{F}) = \mathrm{F} \\ t(\mathrm{F}, \mathrm{F}, \mathrm{F}) = \mathrm{F} \end{cases}$$

is a truth function, viz. conjunction again, but this time the conjunction of *three* items.

The central point to grasp about truth functions is that they are relations between *sequences of truth values* and *truth values. Nothing else matters.* So, for instance, truth-functional relations between propositions are relations *simply between the truth values of the propositions concerned.* The meanings of the propositions enter in only in this respect, that a proposition has to have some meaning if it is to be true or false. But in order to determine whether a proposition stands in a certain truth-functional relation to another, you do not need to know *what* either proposition means: it is enough to know their truth values (and sometimes you do not even need to know what their truth values are: it may be enough to know *that* they have truth values, i.e. that they are propositions, in the sense defined).

*Definition.* A *truth-functional propositional connective* is a meaningful expression or symbol that can be combined with propositions (or formulas) to form propositions (or formulas) and that can be completely defined by a complete standard truth table (i.e. one in which for each row of the table the final column has a single definite truth value).[1]

Example: The symbol '$\supset$' is a truth-functional propositional connective. It can be completely defined by the following truth table:

| A | B | A $\supset$ B |
|---|---|---|
| T | T | T |
| F | T | T |
| T | F | F |
| F | F | T |

(The order in which the four rows are written is unimportant.)

Each truth-functional propositional connective corresponds to a unique truth function, in the way now illustrated:

To the connective '$\supset$' there corresponds the function

$$\begin{cases} q(\mathrm{T}, \mathrm{T}) = \mathrm{T} \\ q(\mathrm{F}, \mathrm{T}) = \mathrm{T} \\ q(\mathrm{T}, \mathrm{F}) = \mathrm{F} \\ q(\mathrm{F}, \mathrm{F}) = \mathrm{T} \end{cases}$$

i.e. material implication.

(The order in which the four rows are written is unimportant.)

A monadic connective is a connective that combines with *one* proposition or formula to form a new one; a dyadic, or binary, connective combines with *two* propositions (formulas) to form a new one; and so on. '$\sim$' is a monadic connective; '$\supset$' and '$\wedge$' and '$\vee$' are dyadic (binary) connectives.[2] There are no familiar triadic connectives, since (as we shall show) everything that can be expressed by means of triadic or more complicated connectives can be expressed using only dyadic connectives.

There are $2^2 = 4$ distinct truth functions of one argument, viz.

$$\begin{cases} f_1(\mathrm{T}) = \mathrm{T} \\ f_1(\mathrm{F}) = \mathrm{T} \end{cases} \qquad \text{[No name]}$$

[1] See exercise 2 at the end of the section for examples of connectives that can and connectives that cannot be so defined.

[2] '$\wedge$' is a symbol for conjunction, '$\vee$' for [inclusive] disjunction.

$$\begin{cases} f_2(\mathrm{T}) = \mathrm{T} \\ f_2(\mathrm{F}) = \mathrm{F} \end{cases} \quad \text{[Identity]}$$

$$\begin{cases} f_3(\mathrm{T}) = \mathrm{F} \\ f_3(\mathrm{F}) = \mathrm{T} \end{cases} \quad \text{[Negation]}$$

$$\begin{cases} f_4(\mathrm{T}) = \mathrm{F} \\ f_4(\mathrm{F}) = \mathrm{F} \end{cases} \quad \text{[No name]}$$

There are $(2^2)^2 = 16$ truth functions of two arguments, e.g.

$$\begin{cases} g_1(\mathrm{T}, \mathrm{T}) = \mathrm{T} \\ g_1(\mathrm{F}, \mathrm{T}) = \mathrm{T} \\ g_1(\mathrm{T}, \mathrm{F}) = \mathrm{T} \\ g_1(\mathrm{F}, \mathrm{F}) = \mathrm{T} \end{cases}$$

$$\begin{cases} g_2(\mathrm{T}, \mathrm{T}) = \mathrm{T} \\ g_2(\mathrm{F}, \mathrm{T}) = \mathrm{T} \\ g_2(\mathrm{T}, \mathrm{F}) = \mathrm{T} \\ g_2(\mathrm{F}, \mathrm{F}) = \mathrm{F} \end{cases} \qquad \text{etc.}$$

There are $((2^2)^2)^2 = 16^2 = 256$ truth functions of three arguments. There are $2^{(2^m)}$ truth functions of $m$ arguments.

For each positive integer $n$ there is a finite number of distinct truth functions of $n$ arguments. So the set of all truth functions is denumerable.

Do we need denumerably many connectives to express the denumerably many truth functions? We shall see later that the answer is 'No'. They can all be expressed by means of a single dyadic connective.

### EXERCISES

1. Give a rule defining the truth function of three arguments that has the value truth when the second argument has the value truth, and the value falsity otherwise.

2. Which of the following are (in their context) truth-functional propositional connectives?

(*a*) 'It is not the case that' in the sentence 'It is not the case that Napoleon won the Battle of Waterloo'.

(*b*) 'And' in '2 + 2 = 4 and Napoleon won the Battle of Waterloo'.

(*c*) 'And then' in 'He took off his clothes and then he jumped into the water'.

(*d*) 'Hunter believes that' in 'Hunter believes that Napoleon won the Battle of Waterloo'.
(*e*) 'If' in 'If Tom marries Mary, Susan will be unhappy'.
(*f*) 'Either . . . or . . .' in '2 + 2 = 4, so either 2 + 2 = 4 or there is life on Mars'.
(*g*) 'Either . . . or . . .' in 'Either he caught the bus or he had to walk'.
(*h*) 'If' in 'If he's a millionaire, I'm a Dutchman'.

ANSWERS

1. Let the function be $g$. Then the rule is

$$\begin{cases} g(\mathrm{T}, \mathrm{T}, \mathrm{T}) = \mathrm{T} \\ g(\mathrm{F}, \mathrm{T}, \mathrm{T}) = \mathrm{T} \\ g(\mathrm{T}, \mathrm{F}, \mathrm{T}) = \mathrm{F} \\ g(\mathrm{F}, \mathrm{F}, \mathrm{T}) = \mathrm{F} \\ g(\mathrm{T}, \mathrm{T}, \mathrm{F}) = \mathrm{T} \\ g(\mathrm{F}, \mathrm{T}, \mathrm{F}) = \mathrm{T} \\ g(\mathrm{T}, \mathrm{F}, \mathrm{F}) = \mathrm{F} \\ g(\mathrm{F}, \mathrm{F}, \mathrm{F}) = \mathrm{F} \end{cases}$$

2. The connectives in (*a*), (*b*) and (*f*) are truth-functional propositional connectives. The others are not.

(*a*) 'It is not the case that' can be completely defined by the truth table

| A | It is not the case that A |
|---|---|
| T | F |
| F | T |

(*b*) This 'and' can be completely defined by the truth table

| A | B | A and B |
|---|---|---|
| T | T | T |
| F | T | F |
| T | F | F |
| F | F | F |

(*c*) If we try to construct a truth table for 'and then' we find that the truth value for one row is left undetermined:

| A | B | A and then B |
|---|---|---|
| T | T | ? |
| F | T | F |
| T | F | F |
| F | F | F |

(*d*) Here the values for both rows are left undetermined:

| A | Hunter believes that A |
|---|---|
| T | ? |
| F | ? |

(*e*) Values for three rows left undetermined:

| A | B | If A then B |
|---|---|---|
| T | T | ? |
| F | T | ? |
| T | F | F |
| F | F | ? |

(*f*) This truth-functional (or 'extensional') use of 'either . . . or . . .' can be completely defined by the truth table

| A | B | Either A or B |
|---|---|---|
| T | T | T |
| F | T | T |
| T | F | T |
| F | F | F |

(*g*) This is a non-truth-functional (or 'intensional') use of 'either . . . or . . .'. The whole sentence is equivalent to 'If he did not catch the bus, then he had to walk', and, just as in Case (*e*), the values for three of the rows are left undetermined:

| A | B | Either A or B | = | If not A, then B |
|---|---|---|---|---|
| T | T | ? | | ? |
| F | T | ? | | ? |
| T | F | ? | | ? |
| F | F | F | | F |

See further Strawson (1952, p. 90).

(*h*) As for (*e*). Cf. Strawson (1952, p. 89).

## 17 A formal language for truth-functional propositional logic: the formal language P

We define now a formal language that on its intended interpretation will be capable of expressing truths of truth-functional propositional logic. But our definition will make no essential reference to this or any interpretation. The terms 'propositional symbol', 'connective', 'bracket' used in describing the language *are therefore to be taken purely as handy labels*, devised with an eye on the intended interpretation, certainly, but replaceable in the context by arbitrary conglomerations of letters, such as 'schlumpf', 'snodgrass' or 'zbolg'. We shall call the language 'P' (for 'propositional logic').

### *The formal language P*

*Symbols of P*

P has exactly six symbols, viz.:

$p$
$'$
$\sim$
$\supset$
$($
$)$

Names for these symbols:

The symbol $p$
The dash
The tilde
The hook
The left-hand bracket
The right-hand bracket

We shall call the tilde and the hook the *connectives* of P.

We shall say that the symbol $p$ followed by one or more

dashes is a *propositional symbol* of P. So each of the following is a propositional symbol of P:

$$p'$$
$$p''$$
$$p'''$$
$$p''''$$

*Formulas (wffs) of P*

1. Any propositional symbol is a wff of P.
2. If A is a wff of P, then the string of symbols of P consisting of the tilde followed by the formula A is a wff of P. (We abbreviate this to: If A is a wff of P, then $\sim$A is a wff of P.)
3. If A and B are wffs of P, then the string of symbols of P consisting of the left-hand bracket, the formula A, the hook, the formula B, and the right-hand bracket, in that order, is a wff of P. (Abbreviated to: If A and B are wffs of P, then (A $\supset$ B) is a wff of P.)
4. Nothing else is a wff of P.

[In this description the letters 'A' and 'B' are metalinguistic variables.]

*Examples*: The following are wffs of P:

$$p'''''$$
$$\sim p''$$
$$(p''' \supset p')$$
$$\sim(p''' \supset p')$$
$$(\sim\sim(p''' \supset p') \supset \sim p'')$$

The following are *not* wffs of P:

| | |
|---|---|
| $p$ | [No dash] |
| $\sim(p'')$ | [Superfluous brackets] |
| $(\sim p'')$ | [Superfluous brackets] |
| $p''' \supset p'$ | [No brackets] |
| $q''$ | ['*q*' is not a symbol of P] |
| $(p \supset q)$ | [Obvious] |
| (A $\supset$ B) | ['A' and 'B' are not symbols of P] |

## 18 Conventions: 1. About quotation marks 2. About dropping brackets

If we were to adhere strictly to our own requirements about the use of quotation marks and brackets, the rest of this book would be even less readable than it now is.[1] Accordingly from now on we adopt the following conventions:

1. *About quotation marks*

Each symbol or formula is to be regarded as a name or description of itself, if the context so requires.

E.g. instead of writing

$$\text{`}(p' \supset p')\text{'} \text{ is a formula of P}$$

we shall write simply

$$(p' \supset p') \text{ is a formula of P}$$

and instead of writing

$$\text{`}p\text{'} \text{ is a symbol}$$

we shall write simply

$$p \text{ is a symbol.}$$

2. *About dropping brackets*

From now on we shall usually drop the outermost pair of brackets from a formula. E.g. instead of writing

$$(p' \supset (p'' \supset p'))$$

we shall write simply

$$p' \supset (p'' \supset p').$$

*Note*. We shall be flexible in our use of these conventions, sometimes putting in the full supply of quotation marks or brackets, when that seems the natural thing to do in the context.

[1] Cf. Hobbes: 'Your Treatise of the *Angle of Contact*, I have before confuted in a very few leaves. And for that of your *Conique Sections*, it is so covered over with the scab of Symboles that I had not the patience to examine whether it be well or ill demonstrated.' (*Six Lessons to the Professors of the Mathematiques . . . in the University of Oxford*, Lesson V, p. 49 [1656]; in *The English Works of Thomas Hobbes*, ed. Sir William Molesworth, vol. VII (1845), where see p. 316.)

## 19 Semantics for P. Definitions of *interpretation of P*, *true/false for an interpretation of P*, *model of a formula/set of formulas of P*, *logically valid formula of P*, *model-theoretically consistent formula/set of formulas of P*, *semantic consequence* (for formulas of P), *tautology of P*

This section is mostly just an abstract statement of what is normally explained by means of the usual truth tables.

*Definition.* An *interpretation of P* is an assignment to each propositional symbol of P of one or other (but not both) of the truth values truth and falsity, and an assignment to the connectives of P of their usual truth-functional meanings (which we shall define more precisely in clauses 2 and 3 of the definition of *true for an interpretation of P*, below).

For $n$ distinct propositional symbols there are $2^n$ distinct possible interpretations. For the symbol $p'$, for example, there are $2^1 = 2$ possible interpretations, viz.

(1) $p'$ is assigned T
(2) $p'$ is assigned F.

For the pair $p'$, $p''$ there are $2^2 = 4$ possible interpretations, viz.

(1) both assigned T
(2) $p'$ assigned F, $p''$ assigned T
(3) $p'$ assigned T, $p''$ assigned F
(4) both assigned F.

Since P has $\aleph_0$ (i.e. denumerably many) propositional symbols, there are $2^{\aleph_0} = \mathfrak{c}$ (and so uncountably many) distinct possible interpretations of P. (For $\aleph_0$ and $\mathfrak{c}$, see Appendix 1.)

*Definition* of *true for an interpretation of P*

Let I be any interpretation of P, and A and B any formulas of P. Then:

1. If A is a propositional symbol, then A is true for I iff I assigns the truth value truth to A.
2. $\sim$A is true for I iff A is not true for I.
3. (A $\supset$ B) is true for I iff either A is not true for I or B is true for I.[1]

[1] The 'either . . . or . . .' in this sentence is intended to be purely truth-functional. Clause 3 could be put in the form '. . . iff not both A is true for I and B is not true for I'.

*Definition* of *false for an interpretation of P*

A is false for I iff A is not true for I.

When a formula is true/false for a given interpretation we shall say that it has the truth value truth/falsity for that interpretation.

*Definition.* An interpretation I is a *model* of a formula (alternatively: set of formulas) of P iff the formula (alternatively: every formula in the set) is true for I.

Until further notice, in what follows A, B, C, etc., are to be arbitrary formulas of P.

*Definition.* A is a *logically valid* formula of P [$\models_P$] iff A is true for every interpretation of P.

Examples: The following wffs are logically valid formulas of P (remember the convention about dropping brackets):

$$p' \supset p'$$
$$p' \supset (p'' \supset p')$$
$$(p' \supset (p'' \supset p''')) \supset ((p' \supset p'') \supset (p' \supset p'''))$$
$$(\sim p' \supset \sim p'') \supset (p'' \supset p')$$

The following are *not* logically valid wffs of P:

$$p'$$
$$\sim p'$$
$$\sim (p' \supset p')$$

'$\models_P$' is a symbol of the metalanguage, not of the object language P itself. Thus

$$\models_P p'$$

is an abbreviation for

'$p'$' is a logically valid formula of P.

With both the symbol '$\models$' and the symbol '$\vdash$' [§22 below] it is the convention to omit any quotation marks that it would be natural to supply. E.g. we write

$$\models_P p'$$

and not

$$\models_P \text{'}p'\text{'}.$$

It is not usual to put a subscript on '$\vDash$', as we have done. Usually there is only one formal language being discussed. However, since we shall be discussing several different formal languages, to avoid confusion we shall on occasion put subscripts on '$\vDash$' to indicate which formal language we are talking about.

*Definition.* A *model-theoretically consistent* (*m-consistent*) formula or set of formulas of P is one that has a model.

A *model-theoretically inconsistent* (*m-inconsistent*) formula or set of formulas of P is one that has no model.

Later we shall define the notions of *proof-theoretically consistent/inconsistent* formula or set of formulas of a formal system PS. It has to be proved that a formula or set of formulas of P is model-theoretically consistent iff it is a proof-theoretically consistent formula or set of formulas of PS. We do this via the Henkin proof of the semantic completeness of PS, §32. [For the system PS, see §22.]

*Definition. A formula B of P is a semantic consequence of a formula A of P* [$A \vDash_P B$] iff there is no interpretation of P for which A is true and B is false.[1]

(So if there is no interpretation of P for which A is true, i.e. if A is an m-inconsistent formula, then any formula of P you like to take is a semantic consequence of A.)

*Definition. B is a semantic consequence of a set Γ of formulas of P* [$\Gamma \vDash_P B$] iff there is no interpretation of P for which every formula in Γ is true and B is false.[1]

*The empty set*: ∅

We adopt the usual convention that every interpretation of P is a model of the empty set. Then:

19.1 $\emptyset \vDash_P A$ *iff* $\vDash_P A$

i.e. A is a semantic consequence of the empty set iff A is logically valid.

*Consequences of these definitions*

19.2 *For any given interpretation a given formula is either true or false*

[1] The definition of semantic consequence for languages adequate for predicate logic is more complicated. See Part 3, §39.

19.3 *No formula is both true and false for the same interpretation*

19.4 *A is false for a given interpretation iff* $\sim A$ *is true for that interpretation; and A is true for an interpretation iff* $\sim A$ *is false for that interpretation*

19.5 *If A and* $A \supset B$ *are both true for a given interpretation, then B is true for that interpretation*

19.6 *If* $\models_P A$ *and* $\models_P A \supset B$, *then* $\models_P B$

19.7 *B is a semantic consequence of A iff* $A \supset B$ *is logically valid: i.e.* $A \models_P B$ *iff* $\models_P A \supset B$

Proofs of 19.5 and 19.6:

19.5 *If A and* $A \supset B$ *are both true for a given interpretation, then B is true for that interpretation*

*Proof.* Suppose that for some interpretation A and $A \supset B$ are both true. Then by clause 3 of the definition of *true for an interpretation of P* on p. 57 B is also true for that interpretation.

19.6 *If* $\models_P A$ *and* $\models_P A \supset B$, *then* $\models_P B$

*Proof.* Assume A and $A \supset B$ are both logically valid while B is not. Then for some interpretation B is not true. For that interpretation A will be true and $A \supset B$ false. But this contradicts our assumption that $A \supset B$ was logically valid. Therefore if A and $A \supset B$ are both logically valid, so is B.

*Definition.* A is a *tautology of P* iff A is true for every assignment of truth values to its propositional symbols when the connectives bear their usual truth-table meanings: i.e. iff A is true for every interpretation of P: i.e. iff A is a logically valid formula of P.

(In the case of the language P we need not, indeed cannot, distinguish tautologies of P from logically valid formulas of P. Later, when we come to the language Q, adequate for predicate logic, we shall find that the tautologies of Q are a proper subset of the logically valid formulas of Q. Here we are just preparing the ground.)

Intuitively, a tautology is a formula that can be checked to be true for all interpretations by the usual (finite) truth-table method. Some logically valid formulas of the predicate language Q cannot be so checked.

## 20 Some truths about $\vDash_P$. The Interpolation Theorem for P

Where A, B are arbitrary formulas of P and Γ, Δ arbitrary sets of formulas of P:

20.1 $A \vDash_P A$

20.2 *If* $\Gamma \vDash_P A$, *then* $\Gamma \cup \Delta \vDash_P A$

20.3 *If* $\Gamma \vDash_P A$ *and* $A \vDash_P B$, *then* $\Gamma \vDash_P B$

20.4 *If* $\Gamma \vDash_P A$ *and* $\Gamma \vDash_P A \supset B$, *then* $\Gamma \vDash_P B$

20.5 *If* $\vDash_P A$, *then* $\Gamma \vDash_P A$

These are all more or less immediate consequences of definitions.

For convenience of reference we repeat here 19.6 and 19.7:

19.6 *If* $\vDash_P A$ *and* $\vDash_P A \supset B$, *then* $\vDash_P B$

19.7 $A \vDash_P B$ *iff* $\vDash_P A \supset B$

20.6 (*The Interpolation Theorem for P*) *If* $\vDash_P A \supset B$ *and A and B have at least one propositional symbol in common, then there is a formula C of P all of whose propositional symbols occur in both A and B such that* $\vDash_P A \supset C$ *and* $\vDash_P C \supset B$

*Informal proof*

1. Suppose every propositional symbol in A also occurs in B. Then we just let C be A itself, for obviously if $\vDash_P A \supset B$, then $\vDash_P A \supset A$ and $\vDash_P A \supset B$.

2. Now suppose that there is just one propositional symbol that occurs in A but does not occur in B. Call it '*p*'. Since by hypothesis A ⊃ B is logically valid, A ⊃ B takes the value T when *p* is assigned T and it also takes the value T when *p* is assigned F. Let *q* be any propositional symbol that occurs in both A and B. Let $A_1$ be the formula that results from A when $(q \supset q)$ is substituted for *p* in A, and let $A_2$ be the formula that results from A when $\sim(q \supset q)$ is substituted for *p* in A. Then $A_1 \supset B$ and $A_2 \supset B$ are both logically valid ($A_1$ amounts to substituting T for *p* in A; $A_2$ to substituting F for *p* in A). Standard truth-table reasoning will show that, given the definitions of $A_1$ and $A_2$, $A \supset (A_1 \vee A_2)$ is a truth-table tautology, and so, since $A_1 \supset B$ and $A_2 \supset B$ are both tautologies of P (see above),

$(A_1 \vee A_2) \supset B$ is also a truth-table tautology. Now '$\vee$' is not a symbol of P. But for any formulas A and B

$$(A \vee B) \equiv (\sim A \supset B).$$

So we can rewrite $A_1 \vee A_2$ as $\sim A_1 \supset A_2$. Then we have:

If $\models_P A \supset B$, then $\models_P A \supset (\sim A_1 \supset A_2)$ and $\models_P (\sim A_1 \supset A_2) \supset B$.

So we have our C, viz. $(\sim A_1 \supset A_2)$.

3. If there is more than one propositional symbol that occurs in A but not in B, we let $A_1/A_2$ be the formula got from A by replacing *each* one by $(q \supset q)/\sim(q \supset q)$. Then the rest of the argument goes through as before.

A rigorous proof of this theorem, by mathematical induction on the number of propositional symbols in A but not in B, is given as the answer to the exercise on §27.

## 21 P's powers of expression. Adequate sets of connectives

We shall prove (Theorem 21.1) that the language P is capable of expressing any truth function, in the following sense:

> To every truth function there corresponds in a natural way a complete truth table. To every complete truth table there corresponds (not necessarily uniquely) a formula of P ('corresponds' in the sense of having that truth table as its truth table).

Examples:

1. To the truth function *material implication* there corresponds the table

$$\langle T, T\rangle = T$$
$$\langle F, T\rangle = T$$
$$\langle T, F\rangle = F$$
$$\langle F, F\rangle = T$$

To that table there corresponds the formula $p' \supset p''$, among others:

| $p'$ | $p''$ | $p' \supset p''$ |
|---|---|---|
| T | T | T |
| F | T | T |
| T | F | F |
| F | F | T |

2. To the nameless truth function to which there corresponds the table

$$\langle T, T, T\rangle = F$$
$$\langle F, T, T\rangle = F$$
$$\langle T, F, T\rangle = F$$
$$\langle F, F, T\rangle = F$$
$$\langle T, T, F\rangle = F$$
$$\langle F, T, F\rangle = T$$
$$\langle T, F, F\rangle = F$$
$$\langle F, F, F\rangle = F$$

there corresponds (among others) the formula

$$\sim(\sim(\sim p' \supset \sim p'') \supset \sim \sim p''')$$

thus:

| $p'$ | $p''$ | $p'''$ | $\sim(\sim(\sim p' \supset \sim p'') \supset \sim \sim p''')$ |
|---|---|---|---|
| T | T | T | F |
| F | T | T | F |
| T | F | T | F |
| F | F | T | F |
| T | T | F | F |
| F | T | F | T |
| T | F | F | F |
| F | F | F | F |

Instead of saying that P is capable of expressing any truth function, we shall say that the set of connectives $\{\sim, \supset\}$ is adequate for the expression of any truth function, since the only connectives in P are $\sim$ and $\supset$. Some other sets of connectives are also adequate; i.e. some languages, that differ from P only in having different connectives, are also capable of expressing any truth function. ('Different connectives' here means connectives that differ in their truth-table definition, not merely in the physical shape of their tokens.)

Theorem 21.1 (which we still have to prove) is our first important metatheorem:

21.1 *The set* $\{\sim, \supset\}$ *is adequate for the expression of any truth function* [*so P can express any truth function*]

The proof of Metatheorem 21.1 is in two stages. We prove first that the set $\{\sim, \wedge, \vee\}$ is adequate (Metatheorem 21.2); then that if the set $\{\sim, \wedge, \vee\}$ is adequate, then the set $\{\sim, \supset\}$ is.

21.2 *The set* $\{\sim, \wedge, \vee\}$ *is adequate*

*Proof.* Intuitively, the proof consists in showing that, given any complete truth table, we can construct a formula[1] in disjunctive normal form that has that table as its truth table.

A formula is in *disjunctive normal form* (DNF) iff it is a disjunction of conjunctions of single propositional symbols or their negations; counting as degenerate cases of disjunctions / conjunctions single propositional symbols and their negations, and allowing disjunctions / conjunctions of more than two disjuncts / conjuncts.

Examples (throughout we only include such brackets as are necessary to avoid ambiguity):

1. $(p' \wedge \sim p'' \wedge p''') \vee (p'' \wedge \sim p') \vee (\sim p'''' \wedge p''')$
2. $(p' \wedge \sim p'' \wedge p''') \vee (p'' \wedge \sim p') \vee \sim p''''$
   [$\sim p''''$ counts as a degenerate conjunction with only one conjunct.]
3. $p' \vee p''$
   [Each of $p'$ and $p''$ counts as a degenerate conjunction.]
4. $p' \wedge p''$
   [This counts as a degenerate disjunction with only one disjunct, viz. the whole formula.]
5. $\sim p'$
   [This counts as a degenerate disjunction of a degenerate conjunction.]

Notice that a formula is in DNF only if

(1) the only connectives that occur in it are connectives for negation, conjunction and disjunction (not necessarily all of these), and

[1] From here to the end of the section 'formula' is used to cover not only formulas of P but also formulas of languages with truth-functional propositional connectives additional to those in P.

(2) negation is over single propositional symbols only, not over any more complicated expressions (e.g. not over conjunctions or disjunctions).

Now the proof can go quite simply:

Let $f$ be any arbitrary truth function of an arbitrary number, $n$, of arguments. Write out the complete truth table corresponding to $f$. It will have $n + 1$ columns and $2^n$ rows. Look at the T's and F's in the last column (i.e. the column that gives the values of the function for the sets of arguments in the corresponding rows). There are three possibilities:

1. The last column is all F's.
2. There is exactly one T in the last column.
3. There is more than one T in the last column.

We show in each case how to construct a formula in DNF having $n$ distinct propositional symbols and the same truth table as $f$.

*Case 1* (*The last column is all F's*)

Then

$$p' \wedge \sim p' \wedge p'' \wedge p''' \wedge \ldots \wedge p^n$$

[where $p^n$ is an abbreviation for $p$ followed by $n$ dashes] is a formula in DNF that has the same truth table as $f$. For $p' \wedge \sim p'$ always gets F, and so therefore does anything of which it is a conjunct.

*Case 2* (*The last column has just one T*)

Go along the row that has the T in its final column. If the first entry in the row is T, write $p'$; if the first entry is F, write $\sim p'$. If the second entry is T, write $p''$; if it is F, write $\sim p''$. And so on, as far as and including the $n$th entry. Form the *conjunction* of what you have written (i.e. insert $n - 1$ conjunction signs in the appropriate places). The resulting formula will be in DNF and have the same truth table as the function $f$.

Example: Let $f$ be the function of three arguments that has the table

T T T = F
F T T = F
T F T = T
F F T = F

| | | | |
|---|---|---|---|
| T | T | F | = F |
| F | T | F | = F |
| T | F | F | = F |
| F | F | F | = F |

Then the formula

$$p' \wedge \sim p'' \wedge p'''$$

is in DNF and has the same truth table as *f*. It has the value T iff $p'$ has T, $p''$ has F, and $p'''$ has T; in all other cases it has the value F.

*Case 3* (*More than one T*)

For each row that ends in a T construct a formula as in Case 2. Form the *disjunction* of all these formulas. The resulting formula will be in DNF and have the same truth table as *f*.

Example: Let *f* be the function of three arguments that has the table

| | | | |
|---|---|---|---|
| T | T | T | = F |
| F | T | T | = F |
| T | F | T | = T |
| F | F | T | = F |
| T | T | F | = F |
| F | T | F | = T |
| T | F | F | = F |
| F | F | F | = T |

Then the formula

$$(p' \wedge \sim p'' \wedge p''') \vee (\sim p' \wedge p'' \wedge \sim p''') \vee (\sim p' \wedge \sim p'' \wedge \sim p''')$$

is in DNF and has the same truth table as *f*. It has the value T in each of the three cases

(1) $p'$ T, $p''$ F, $p'''$ T
(2) $p'$ F, $p''$ T, $p'''$ F
(3) $p', p'', p'''$ all F

and the value F otherwise.

This completes the proof of Metatheorem 21.2.

*Proof of Metatheorem* 21.1 (*The set* $\{\sim, \supset\}$ *is adequate*)

1. Any *conjunction* of two formulas A and B has the same truth table as a formula in which A and B are related by $\sim$ and $\supset$ instead of $\wedge$, thus:

$$(A \wedge B) \equiv \sim(A \supset \sim B).$$

Let C be any formula in which $\wedge$ occurs. Then by replacing every subformula of C that has the form $(A \wedge B)$ by a subformula of the form $\sim(A \supset \sim B)$ we get a formula in which $\wedge$ does not occur and that has the same truth table as C.

2. Similarly for $\vee$. Any *disjunction* of two formulas A and B has the same truth table as a formula in which A and B are related by $\sim$ and $\supset$ instead of $\vee$, thus:

$$(A \vee B) \equiv (\sim A \supset B) \quad \text{[i.e. ([}\sim A] \supset B)].$$

3. Let W be any formula in which either $\wedge$ or $\vee$ occurs, or both $\wedge$ and $\vee$ occur. By carrying out successively the replacement operations described in (1) and (2) above, we get a formula W′ in which no connectives other than $\sim$ and $\supset$ occur and that has the same truth table as W.

4. So since the set $\{\sim, \wedge, \vee\}$ is adequate for the expression of any truth function [21.2], so also is the set $\{\sim, \supset\}$.

Q.E.D.

By similar arguments other sets of connectives can also be shown to be adequate.

21.3 *The set* $\{\sim, \vee\}$ *is adequate* [Emil L. Post, 1920]

*Proof.* Use Metatheorem 21.2 and the tautological schema

$$(A \wedge B) \equiv \sim(\sim A \vee \sim B).$$

21.4 *The set* $\{\sim, \wedge\}$ *is adequate*

*Proof.* Use Metatheorem 21.2 and the tautological schema

$$(A \vee B) \equiv \sim(\sim A \wedge \sim B).$$

C. S. Peirce in a paper of about 1880 that he did not publish (‘A Boolian Algebra with One Constant’, *Collected Papers*, iv, §§12–20 [pp. 13–18]) presented a language for Boolean algebra with just one constant ‘which serves at the same time as the only sign for compounding terms and which renders special signs for negation, for “what is” and for “nothing” unnecessary’. For our present purpose we can take it as a dyadic connective meaning ‘Neither A nor B’. Peirce claimed that it was adequate, but he did not give a rigorous proof of its adequacy. Later, in another unpublished paper, written in 1902 (*Collected Papers*, iv, §265 [p. 216]), he showed that anything that could be expressed by the connective meaning 'Neither A nor B' could equally be expressed using only a connective meaning 'Either not A or not B'. Henry M. Sheffer, without knowing Peirce's result, showed (1912) that all truth functions expressible by means of the primitive connectives ($\sim$, $\vee$) of *Principia Mathematica* could be expressed by either of Peirce's two connectives. Emil L. Post was the first to give a completely general proof of adequacy (for $\{\sim, \vee\}$, in his doctoral dissertation for Columbia University, completed in 1920 and published in the following year: cf. Post, 1920).

21.5 *The set* $\{\downarrow\}$ *is adequate* [C. S. Peirce, *c.* 1880; H. M. Sheffer, 1912. But see comment above.]

A $\downarrow$ B has the value T iff A and B both have the value F. So $p \downarrow q$ can be read as 'Neither $p$ nor $q$'.

*Proof.* Use Metatheorem 21.4 and the tautological schemata

$$\sim A \equiv A \downarrow A, \quad (A \wedge B) \equiv (A \downarrow A) \downarrow (B \downarrow B).$$

21.6 *The set* $\{|\}$ *is adequate* [C. S. Peirce, 1902; H. M. Sheffer, 1912. But see comment preceding 21.5.]

The symbol | expresses what is usually called 'the Sheffer stroke function' (for which see the comment preceding 21.5). A|B has the value F iff A and B both have the value T. So $p|q$ can be read as 'Not both $p$ and $q$' or as 'Either not $p$, or not $q$, or not $p$ and not $q$'.

*Proof.* Use Metatheorem 21.3 and the tautological schemata

$$\sim A \equiv A|A, \quad (A \vee B) \equiv (A|A)|(B|B).$$

There are other adequate sets, and some inadequate ones.

21.7 *The set* $\{\wedge, \vee\}$ *is adequate*

The proof (which here we only indicate) is by showing that the negation of a formula cannot be expressed by any combination of propositional symbols, $\wedge$, and $\vee$. Let P′ be a language just like P, except that it has the connectives $\wedge$ and $\vee$ in place of the connectives $\sim$ and $\supset$. It is shown that (1) no formula of P′ that consists of just one [occurrence of a] symbol can have the value T when all its component propositional symbols have the value F. Then it is shown that (2) *if* this is true of *every* formula of P′ with fewer than $m$ [occurrences of] symbols, *then* it is also true of every formula of P with exactly $m$ symbols. It follows that *no* formula of P′ can have the value T when all its component propositional symbols have the value F, and therefore that P′ cannot express negation. [This type of proof is known as proof by (strong) mathematical induction, about which more later.]

21.8 *The set* $\{\wedge, \supset\}$ *is inadequate*
Proof similar to that for 21.7.

21.9 *The set* $\{\supset, \vee\}$ *is inadequate*
Proof similar to that for 21.7.

Not every set that has $\sim$ as a member is adequate, and not every set that does not is inadequate:

21.10 *The set* $\{\sim, \equiv\}$ *is inadequate*
The proof is similar to that for 21.7, but in this case we show that material implication, for example (we could equally well take conjunction, or disjunction), cannot be expressed by any combination of propositional symbols, $\sim$, and $\equiv$. For the truth table for material implication has four rows and a final column with three T's and one F; while any four-rowed truth table for any formula with no connectives other than $\sim$ and $\equiv$ must have either all T's in its final column, or all F's, or two T's and two F's. (This is rigorously proved by mathematical induction in the answer to exercise 2 of §27, p. 90.)

21.11 *Material implication and exclusive disjunction together are adequate*
[There is no agreed symbol for exclusive disjunction. The truth table for it is

| A | B | A excl. disj. B |
|---|---|---|
| T | T | F |
| F | T | T |
| T | F | T |
| F | F | F |

It can be seen that exclusive disjunction has the same truth table as negated material equivalence. So we shall use the symbol $\not\equiv$ for exclusive disjunction.]

*Proof.* Use 21.1 and the tautological schema

$$\sim A \equiv (A \not\equiv (A \supset A)).$$

21.12 *The only dyadic connectives that are adequate by themselves are* | *and* ↓ [Żyliński, 1924]

*Proof.* Suppose there was another connective. Let it be $*$. We work out what its truth table would have to be, row by row:

| A | B | A $*$ B |
|---|---|---|
| T | T | ? |
| F | T | ? |
| T | F | ? |
| F | F | ? |

If the entry in the first row were T, then any formula built up using only $*$ would take the value T when all its propositional symbols took the value T. So no combination could express the negation of A. So the entry for the first row must be F. Similarly, the entry for the last row must be T.

This gives us:

| A | B | A $*$ B |
|---|---|---|
| T | T | F |
| F | T | ? |
| T | F | ? |
| F | F | T |

If the second and third entries were both F's, $*$ would have the same table as | (and so would be the same connective as |, in all but the physical shape of its tokens, which is not important from a theoretical point of view).[1] If they were both T's, $*$ would be the same as ↓. That leaves just two possibilities to consider, viz. (1) second entry T, third entry F, and (2) second entry F, third entry T. In the first case we would have

$$A * B \equiv \sim A.$$

In the second case we would have

$$A * B \equiv \sim B.$$

[1] A truth-functional propositional connective is a *meaningful* symbol, not a merely formal symbol.

In either case $*$ would be definable in terms of $\sim$. But $\sim$ is not adequate by itself, because the only functions of one argument definable from it are negation and identity. I.e. starting from a formula A and using only negation we can get formulas that are truth-functionally equivalent to A and formulas that are truth-functionally equivalent to $\sim$A, but nothing else:

$$\begin{array}{l} A \\ \sim A \\ \sim\sim A \\ \sim\sim\sim A \\ \sim\sim\sim\sim A \\ \ldots \end{array}$$

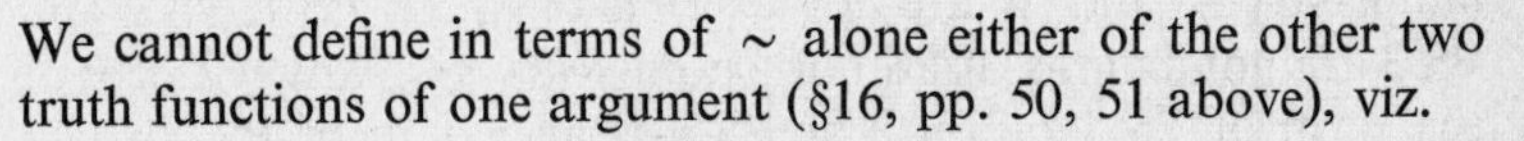

We cannot define in terms of $\sim$ alone either of the other two truth functions of one argument (§16, pp. 50, 51 above), viz.

$$\begin{cases} f_1(T) = T \\ f_1(F) = T \end{cases}$$

and

$$\begin{cases} f_4(T) = F \\ f_4(F) = F \end{cases}$$

So any dyadic connective that is not the same as either $|$ or $\downarrow$ is inadequate by itself.

Q.E.D.

## 22 A deductive apparatus for P: the formal system PS. Definitions of *proof in PS, theorem of PS, derivation in PS, syntactic consequence in PS, proof-theoretically consistent set of PS*

Treat this section as though it followed on directly from the end of §18. Pretend that you know nothing of the contents of §§19–21, i.e. nothing about any interpretation of P.

We now specify a deductive apparatus for the formal language P, viz. a set of axioms and a rule of inference. We call the resulting formal system the system PS (propositional system).

## *The formal system PS*

*Axioms of PS*

If A, B, C are any wffs of P (not necessarily distinct), then the following are axioms of PS (remember the convention about dropping brackets):

[PS 1] $A \supset (B \supset A)$
[PS 2] $(A \supset (B \supset C)) \supset ((A \supset B) \supset (A \supset C))$
[PS 3] $(\sim A \supset \sim B) \supset (B \supset A)$

Nothing else is an axiom of PS.

*Comments*

1. The expressions '$A \supset (B \supset A)$', etc., belong to the metalanguage ('A' and 'B' are not symbols of P). So PS 1, PS 2 and PS 3 are not axioms of PS but axiom-*schemata*. Any wff of P of the form of PS 1 or PS 2 or PS 3 is an axiom of PS. In fact infinitely many formulas of P are axioms of PS in virtue of these three schemata.

Examples: The following are axioms of PS:

| | |
|---|---|
| $p' \supset (p'' \supset p')$ | [by PS 1] |
| $p' \supset (p' \supset p')$ | [by PS 1] |
| $\sim p' \supset (p' \supset \sim p')$ | [by PS 1] |
| $(\sim\sim p' \supset \sim p') \supset (p' \supset \sim p')$ | [by PS 3] |
| $(p' \supset p'') \supset (p' \supset (p' \supset p''))$ | [by PS 1] |

The following are *not* axioms of PS:

| | |
|---|---|
| $A \supset (B \supset A)$ | [not a formula of P] |
| $(A \supset (B \supset C)) \supset ((A \supset B) \supset (A \supset C))$ | [ do. ] |
| $(\sim A \supset \sim B) \supset (B \supset A)$ | [ do. ] |
| $(p' \supset p'') \supset (\sim p'' \supset \sim p')$ | [*not* licensed by PS 3] |

2. There is no question of these axioms being self-evident truths or anything like that. They are merely strings of symbols of the alphabet of P, capable of being interpreted, certainly, but *defined* without reference to any interpretation.

*Rule of inference of PS*

If A and B are any formulas of P, then B is an immediate consequence in PS of the pair of formulas A and $(A \supset B)$.

*Informally*: Given both A and $(A \supset B)$ you may infer B.

*Comment.* In saying that B is an immediate consequence of the *pair* of formulas A and $(A \supset B)$ we mean that you need *both* formulas in order to get B as an immediate consequence. B is not an immediate consequence of A alone or of (A $\supset$ B) alone. The order in which the formulas A and (A $\supset$ B) occur does not matter.

We call this rule 'Modus Ponens for $\supset$', or 'MP' for short.

Examples:

1. $p'$ is an immediate consequence in PS of $p''$ and $(p'' \supset p')$.
2. $p'$ is an immediate consequence in PS of $p'$ and $(p' \supset p')$, but *not* of $p'$ alone.
3. $(p' \supset p')$ is an immediate consequence in PS of $(p'' \supset (p' \supset p'))$ and $p''$.
4. $\sim\sim p'$ is an immediate consequence in PS of $(\sim p' \supset \sim\sim p')$ and $\sim p'$.

*Note.* In stating the rule of inference and giving these examples we have not dropped any brackets.

*Definition.* A *proof in PS* is a finite (but not empty) string of formulas of P each one of which is either an axiom of PS or an immediate consequence by the rule of inference of PS of two formulas preceding it in the string.

Examples: The following are proofs in PS (remember the convention about brackets):

1. [1] $p' \supset ((p' \supset p') \supset p')$ [Axiom, by PS 1]
   [2] $(p' \supset ((p' \supset p') \supset p')) \supset ((p' \supset (p' \supset p')) \supset (p' \supset p'))$ [Axiom, by PS 2]
   [3] $((p' \supset (p' \supset p')) \supset (p' \supset p'))$ [Immediate consequence in PS, by Modus Ponens, of the wffs numbered [1] and [2]]
   [4] $p' \supset (p' \supset p')$ [Axiom, by PS 1]
   [5] $p' \supset p'$ [Immediate consequence, by MP, of [3] and [4]]

*Comment.* The material in square brackets is not part of the proof, but merely explanatory commentary. The proof proper consists simply of the string of five formulas.

2. [1] $p' \supset (p' \supset p')$ [Axiom, by PS 1. This single wff is a complete proof in PS.]
3. [1] $(\sim p'' \supset \sim p') \supset (p' \supset p'')$ [PS 3]
   [2] $((\sim p'' \supset \sim p') \supset (p' \supset p'')) \supset$
   $(\sim p' \supset ((\sim p'' \supset \sim p') \supset (p' \supset p'')))$ [PS 1]

[3] $\sim p' \supset ((\sim p'' \supset \sim p') \supset (p' \supset p''))$ [MP, 1, 2]

[4] $(\sim p' \supset ((\sim p'' \supset \sim p') \supset (p' \supset p''))) \supset$
$((\sim p' \supset (\sim p'' \supset \sim p')) \supset (\sim p' \supset (p' \supset p'')))$ [PS 2]

[5] $(\sim p' \supset (\sim p'' \supset \sim p')) \supset (\sim p' \supset (p' \supset p''))$ [MP, 3, 4]

[6] $\sim p' \supset (\sim p'' \supset \sim p')$ [PS 1]

[7] $\sim p' \supset (p' \supset p'')$ [MP, 6, 5]

*Comments*

1. The formulas mentioned in the definition of *proof in PS* must be complete formulas, not merely subformulas of formulas. E.g. the proof in example 1 is a string of exactly *five* formulas, and the proof in example 2 is a string of exactly *one* formula.

2. According to the definition a proof is a string, and the string could be written out without leaving any gaps between the formulas. E.g. the proof in example 1 could begin like this [putting in the full supply of brackets]:

$$(p' \supset ((p' \supset p') \supset p'))((p' \supset ((p' \supset p')) \supset \ldots$$

The gaps are there simply to make things easier.

*Definition.* A formula A is a *theorem of PS* [$\vdash_{PS}$ A] iff there is some proof in PS whose last formula is A.

Example: $(p' \supset p')$ is a theorem of PS, since there is a proof in PS of which it is the last formula: cf. example 1 on p. 73, ll. 22–9.

'$\vdash$' is a symbol of the metalanguage, not of the object language. The same convention about quotation marks applies with it as with '$\vDash$'.

*Comment.* By our definition every *axiom* of PS will also be a *theorem* of PS. But the converse is not true.

*Definition.* A string of formulas is a *derivation in PS of a wff A from a set Γ of wffs of P* iff

(1) it is a finite (but not empty) string of formulas of P, and
(2) the last formula in the string is A, and
(3) each formula in the string is
either (i) an axiom of PS
or (ii) an immediate consequence by the rule of inference of PS of two formulas preceding it in the string
or (iii) a member of the set Γ.

The set $\Gamma$ may have infinitely many members, finitely many members, or no members at all.

Example: The string

$$\begin{array}{l} p' \\ p' \supset p'' \\ p'' \end{array}$$

is a derivation in PS of the formula $p''$ from the set of formulas of P whose sole members are $p'$ and $p' \supset p''$; i.e. from the set $\{p', p' \supset p''\}$ [quotation marks dropped].

*Comment.* The difference between a *derivation in PS* and a *proof in PS* is this:

In a *proof in PS every* formula is a *theorem* of PS.

In a *derivation in PS* formulas may occur in the string that are *not* theorems of PS; e.g. formulas from $\Gamma$, if $\Gamma$ is a set of formulas that are not theorems of PS. In our example no formula in the derivation is a theorem of PS.

Every *proof in PS* is also a *derivation in PS* of the last formula in the proof (the theorem it proves) *from the empty set*, and also from any other set of formulas of P you like to take. This follows from the definitions of *proof in PS* and *derivation in PS.*

*Definition.* A formula A is a *syntactic consequence in PS* of a set $\Gamma$ of formulas of P [$\Gamma \vdash_{PS} A$] iff there is a derivation in PS of A from the set $\Gamma$.

Example: $p'''$ is a syntactic consequence in PS of the set $\{p', p' \supset p'', p'' \supset p'''\}$, because there is a derivation of $p'''$ from that set, e.g. the derivation

$$\begin{array}{l} p' \\ p' \supset p'' \\ p'' \\ p'' \supset p''' \\ p''' \end{array}$$

Notice that this string is not a proof in PS.

*Comment.* A *derivation* is a string of formulas. A *syntactic consequence* is a formula that stands in a certain relation to a set of formulas.

If there is only one formula in $\Gamma$, it is customary to write $A \vdash_{PS} B$ rather than $\{A\} \vdash_{PS} B$.

*Definition.* A set $\Gamma$ of formulas of P is a *proof-theoretically consistent set of PS* iff for no formula A of P is it the case that both $\Gamma \vdash_{PS} A$ and $\Gamma \vdash_{PS} \sim A$. A set $\Gamma$ is a *proof-theoretically inconsistent set of PS* iff for some formula A of P both $\Gamma \vdash_{PS} A$ and $\Gamma \vdash_{PS} \sim A$.

The notions of a proof-theoretically consistent/inconsistent *formula* of PS are defined in a similar way.

What we call a proof-theoretically consistent set is usually called simply a consistent set. But since we already have the concept of a model-theoretically consistent set and want to distinguish the two sorts of consistency, we use the most natural term for doing so. We abbreviate 'proof-theoretically consistent' to 'p-consistent'. Later we shall show that a set of formulas is a proof-theoretically consistent set of PS iff it is a model-theoretically consistent set of P. But this needs a fairly elaborate proof (§32).

*Comments*

1. A set $\Gamma$ of formulas of P may be a p-inconsistent set of PS even though for no formula A are both A and $\sim A$ *members* of $\Gamma$. p-consistent and p-inconsistent sets are defined in terms of what can be *derived* from them with the help of the axioms and/or rules of inference of PS, and a formula can be derivable in PS from a set of formulas without being a member of the set.

2. The definition of p-consistent set of PS makes essential reference to the *deductive apparatus* of PS. Contrast the definition of m-consistent set, which does not.

3. The notion of a p-consistent set can be similarly defined for other formal systems. A set of formulas that is a p-consistent set of a formal system S may be a p-inconsistent set of another formal system S′ with the same formal language as S. *It all depends on the deductive apparatuses of S and S′*. By contrast an m-consistent set of formulas of a formal language L remains an m-consistent set of formulas of L [1] *no matter what deductive apparatus is added to L.*

[1] Assuming that the notions of *interpretation of L* and *true for an interpretation of L* are held constant.

## 23 Some truths about $\vdash_{PS}$

Where A, B are arbitrary formulas of P and Γ, Δ arbitrary sets of formulas of P:

23.1 $A \vdash_{PS} A$

23.2 *If* $\Gamma \vdash_{PS} A$, *then* $\Gamma \cup \Delta \vdash_{PS} A$

23.3 *If* $\Gamma \vdash_{PS} A$ *and* $A \vdash_{PS} B$, *then* $\Gamma \vdash_{PS} B$

23.4 *If* $\Gamma \vdash_{PS} A$ *and* $\Gamma \vdash_{PS} A \supset B$, *then* $\Gamma \vdash_{PS} B$
[We shall appeal to this one often.]

23.5 *If* $\vdash_{PS} A$, *then* $\Gamma \vdash_{PS} A$

23.6 $\vdash_{PS} A$ *iff* $\emptyset \vdash_{PS} A$

These are exact analogues of the truths about $\models_P$ 20.1–20.5 and 19.1, and again they are all more or less direct consequences of definitions.

23.7 $\Gamma \vdash_{PS} A$ *iff there is a finite subset* $\Delta$ *of* $\Gamma$ *such that* $\Delta \vdash_{PS} A$
This metatheorem follows from our requirement that a derivation must be a *finite* string of formulas.

Metatheorem 23.7 has an exact model-theoretic analogue, viz.

[32.18] $\Gamma \models_P A$ *iff there is a finite subset* $\Delta$ *of* $\Gamma$ *such that* $\Delta \models_P A$

But though we can prove right now that if there is a finite subset Δ of Γ such that $\Delta \models_P A$, then $\Gamma \models_P A$,[1] the proof of the converse will not be given until after the proof of the semantic completeness of PS.

Later, after we have proved the Deduction Theorem for PS (§26), we shall have:

[26.2] $A \vdash_{PS} B$ *iff* $\vdash_{PS} A \supset B$
This is the proof-theoretic analogue of 19.7.

Following custom we write $\Gamma, \Delta \vdash_{PS} A$ instead of $\Gamma \cup \Delta \vdash_{PS} A$. Similarly we write $\Gamma, A \vdash_{PS} B$ instead of $\Gamma \cup \{A\} \vdash_{PS} B$, and $\Gamma, A, B \vdash_{PS} C$ instead of $\Gamma \cup \{A\} \cup \{B\} \vdash_{PS} C$.

[1] Thus: Since Δ is a subset of Γ, $\Gamma = \Gamma \cup \Delta$. So all we have to show is that if $\Delta \models_P A$ then $\Gamma \cup \Delta \models_P A$. But this is simply a notational variant of 20.2 ['If $\Gamma \models_P A$, then $\Gamma \cup \Delta \models_P A$'].

## 24 Concepts of consistency

Alonzo Church (1956, p. 108) writes:

> The notion of *consistency* of a logistic system is semantical in motivation, arising from the requirement that nothing which is logically absurd or self-contradictory in meaning shall be a theorem, or that there shall not be two theorems of which one is the negation of the other. But we seek to modify this originally semantical notion in such a way as to make it syntactical in character (and therefore applicable to a logistic system independently of the interpretation adopted for it) . . .

[There follows a most careful account of various concepts of consistency.]

*Simple consistency*

A system S is *simply consistent* iff for no formula A of S are both A and the negation of A theorems of S.

[For particular systems this definition can be transformed into a purely syntactic (proof-theoretic) one by framing it in terms of the symbolism used in the system to express negation, but without referring to the intended interpretation.]

*Absolute consistency*

A system S is *absolutely consistent* iff at least one formula of S is not a theorem of S.

24.1 *If S is a formal system in which for each formula A of S there is a formula A′ of S that on the intended interpretation expresses the negation of A, then if S is simply consistent then it is absolutely consistent*

*Proof.* Let S be any arbitrary formal system that satisfies the hypothesis of the theorem. Suppose that S is simply consistent. Then there is no formula A of S such that both A and A′ are theorems of S. So for some particular formula B of S either B is not a theorem of S or B′ is not a theorem of S. But both B and B′ are formulas of S. So S is absolutely consistent.

24.2 *If S is a formal system for which it is a metatheorem that* $A, A' \vdash_S B$ *(where A and B are arbitrary formulas of S, and A′*

*is as in* 24.1), *then if S is absolutely consistent then it is simply consistent*

*Proof.* Let S be any arbitrary formal system that satisfies the hypothesis of the theorem. Suppose S is not simply consistent. Then $\vdash_S A$ and $\vdash_S A'$ for some formula A. So, in virtue of the metatheorem mentioned in the hypothesis, $\vdash_S B$ for *any* arbitrary formula B; i.e. every formula of S is a theorem of S. So S is not absolutely consistent. Therefore if S is absolutely consistent then it is simply consistent.

*Note.* There are formal systems of standard truth-functional propositional logic for which it is not a metatheorem that A, $A' \vdash B$ for arbitrary formulas A and B. E.g. it is not a metatheorem of Hiż's formalisation of the standard propositional calculus, described on page 118 below, and we could add to his system as axioms some particular formula and its negation without necessarily making every formula a theorem. That is, we could construct a system that contained the full standard propositional calculus (in the sense of having all the appropriate theorems) and that was absolutely consistent without being simply consistent.

## 25 Proof of the consistency of PS

(*a*) *Proof of the simple and absolute consistency of PS by model-theoretic means*

*Outline of proof.* We give an interpretation for P, and show that on it all the theorems of PS come out true. So we have: If a formula A is a theorem of PS, then it is true for the interpretation, and therefore, by clause 2 of our definition of *true for an interpretation of P*, $\sim A$ is not true for the interpretation, and therefore $\sim A$ is not a theorem of PS. That is, for any formula A of P, if A is a theorem of PS, then $\sim A$ is not a theorem; i.e. PS is simply consistent. To prove the absolute consistency of PS, all we have to do is exhibit a formula of P that is not true for the interpretation and therefore not a theorem of PS.

*Details*

1. Our interpretation for P is this: To every propositional symbol we assign the truth value T. [$\sim$ and $\supset$ are interpreted

as in clauses 2 and 3 of the definition of *true for an interpretation of P* in §19.]

2. *Every axiom of PS is true for this interpretation.*

*Sketch of proof*: It follows from our earlier definitions that any arbitrary formula of P is either true for our interpretation or false for it. So, though each of the three axiom-schemata of PS represents an infinite set of axioms, the A's, B's and C's in those schemata stand for formulas that can only be either true or false on our interpretation, and so, in order to check whether any of the [infinitely many] axioms could be false on our interpretation, all we have to consider is a finite number of possibilities for each axiom-schema, viz. all the possible true–false combinations for its constituent schematic letters. For example, in the case of the axiom-schema PS 1 [$A \supset (B \supset A)$] we have to consider just four possibilities, viz. the cases

(1) where A is true and B is true;
(2) where A is false and B is true;
(3) where A is true and B is false;
(4) where A is false and B is false.

Reference to clause 3 of the definition of *true or an interpretation of P* will show that in each of these four cases $A \supset (B \supset A)$ comes out true; i.e. any formula of P that is an axiom of PS in virtue of axiom-schema PS 1 will be true for our interpretation. We leave it to the reader to check that the same is true for the schemata PS 2 [eight cases to be considered] and PS 3 [four cases].

3. *The (only) rule of inference of PS preserves truth-for-our-interpretation* [i.e. for any arbitrary pair of formulas A and B, if A and $A \supset B$ are both true for our interpretation, then B is also true for our interpretation].

*Proof.* By 19.5.

4. *Every theorem of PS is true for our interpretation.*

*Proof.* Directly from 2 and 3.

5. *PS is simply consistent*: i.e. for no formula A of P are both A and $\sim A$ theorems of PS.

*Proof.* As in the *Outline*, using 4.

6. *PS is absolutely consistent.*

*Proof.* The formula $\sim p'$ (e.g.) is not true for our interpretation. Therefore, by 4, it is not a theorem of PS.

*Summary of this proof*

1. Every axiom of PS is true for the interpretation.
2. The rule of inference of PS preserves truth-for-the-interpretation.
3. Therefore every theorem of PS is true for the interpretation.
4. For no formula A of P are both A and ~A true for the interpretation.
5. Therefore for no formula A of P are both A and ~A theorems of PS: i.e. PS is simply consistent.
6. There is at least one formula of P that is not true for the interpretation: e.g. $\sim p'$.
7. Therefore there is at least one formula of P that is not a theorem of PS: i.e. PS is absolutely consistent.

A different but very similar proof of the simple and absolute consistency of PS is given in §28: cf. the comment after 28.3.

(*b*) *Proof of the simple and absolute consistency of PS by proof-theoretic means*

*Outline of proof.* We define proof-theoretically (in purely syntactic terms) a certain property, X, which, if it belongs to a formula A, does not belong to ~A. We then show that every theorem of PS has the property X. This is enough to show that PS is simply consistent [if A is a theorem of PS, then ~A is not]. To prove absolute consistency we exhibit a formula of P that does not have the property X and therefore is not a theorem of PS.

*Details*

1. For the property X we take a property that we shall call 'syntactic tautologyhood'. The normal logical notion of tautology is a semantic notion, having to do with truth and falsity and so with the interpretation of formulas. To construct the notion of a syntactic tautology we take the normal notion of tautology and empty out of it all its semantic elements. Thus instead of talking about the truth values T and F, we talk about (uninterpreted) symbols '1' and '0' (or any other symbols you like to take). E.g. instead of saying that $p' \supset p'$ is a tautology because its truth table has nothing but T's in its final column, we say that it is a syntactic tautology because it has the (syntactic)

property that its associated 1–0 table [the syntactic analogue of a truth table] has nothing but '1's in its final column.

From the last paragraph it can be seen that for our syntactic proof we need to give syntactic analogues of truth tables for the connectives of P. Here they are:

Table for $\sim$:

| A | $\sim$A |
|---|---|
| 1 | 0 |
| 0 | 1 |

Table for $\supset$:

| A | B | A $\supset$ B |
|---|---|---|
| 1 | 1 | 1 |
| 0 | 1 | 1 |
| 1 | 0 | 0 |
| 0 | 0 | 1 |

We now define an *association to P* [this is just an invented term, not used elsewhere] as an association to each propositional symbol of P of one or other (but not both) of the symbols '1' and '0', together with an association to every other formula of P of '1' or '0' in accordance with the tables for $\sim$ and $\supset$ given above. Finally we define a *syntactic tautology* as a formula that is associated with '1' for *every* association to P.

2. *Every axiom of PS is a syntactic tautology.*

*Proof.* For any association to P every formula of P will be associated with one or other, but not both, of the symbols '1' and '0'. So for each axiom-schema of PS we need only consider a finite number of possibilities [compare the similar argument in stage 2 of the proof by model-theoretic means]. For example, for axiom-schema PS 1 [A $\supset$ (B $\supset$ A)] we only have to consider the four possibilities:

(1) A is associated with '1' and B with '1';
(2) A is associated with '0' and B with '1';
(3) A is associated with '1' and B with '0';
(4) A is associated with '0' and B with '0'.

In each case A $\supset$ (B $\supset$ A) is associated with the symbol '1'. So any formula of P that is an axiom by PS 1 will be associated with '1' for *every* association to P; i.e. it will be a syntactic tautology. Similarly for the other two axiom-schemata.

3. *The (only) rule of inference of PS preserves syntactic tautologyhood.*

*Proof.* Suppose that, for two formulas A and B, A and A ⊃ B are both syntactic tautologies while B is not. There will be a 1–0 table showing the 1–0 associations to A, to B, and to A ⊃ B for each possible combination of 1–0 associations to their constituent propositional symbols. For at least one row in that table B will be associated with '0', by hypothesis. But for that row A will be associated with '1', since it is always associated with '1', by hypothesis. So in that row A ⊃ B will be associated with '0', by the table for ⊃ given in 1 above. But this contradicts our supposition that A ⊃ B is a syntactic tautology. So if A and A ⊃ B are both syntactic tautologies, B must be one too.

4. *Every theorem of PS is a syntactic tautology.*

*Proof.* Directly from 2 and 3.

5. *PS is simply consistent.*

*Proof.* From the table for ~ given in 1 it can be seen that for any formula A, if A is a syntactic tautology, ~A is not. So (from 4) if A is a theorem of PS, ~A is not.

6. *PS is absolutely consistent.*

*Proof.* The formula $p'$ (e.g.) is not a syntactic tautology. Therefore, by 4, it is not a theorem of PS.

*Historical note*

Emil Post (1920) was the first to give a proof of the consistency of the standard propositional calculus. His proof was semantic in character: he showed that all theorems of the system he was dealing with (the propositional system of *Principia Mathematica*) are tautologies. Jan Łukasiewicz seems to have been the first to find a purely syntactic proof, of the kind sketched in (*b*) above. The key idea common to both these consistency proofs (viz. the idea of showing that there is some property that belongs to every axiom, is preserved by the rule(s) of inference, and does not belong to any contradictory formula) goes back to David Hilbert: cf. Hilbert (1904).

## 26 The Deduction Theorem for PS

The Deduction Theorem is a metatheorem that is rather useful for proving other metatheorems. It seems to have been first proved by Alfred Tarski in 1921 [cf. Tarski, 1923–38, p. 32], but the earliest published proof (for a system of predicate logic: the result includes that for a system of propositional logic) was by Jacques Herbrand in 1930 [cf. Herbrand, 1929, pub. 1930]. Herbrand, who made outstanding contributions to mathematical logic, was killed in a mountaineering accident in 1931, at the age of twenty-three.

26.1 (*The Deduction Theorem for PS*) *If* $\Gamma, A \vdash_{PS} B$, *then* $\Gamma \vdash_{PS} A \supset B$

In other words: If B is a syntactic consequence in PS of the set $\Gamma \cup \{A\}$, then $A \supset B$ is a syntactic consequence in PS of the set $\Gamma$ alone.

This in turn means: If there is a derivation in PS of B from $\Gamma \cup \{A\}$, then there is a derivation in PS of $A \supset B$ from $\Gamma$ alone.

Examples: The Deduction Theorem implies that:

1. If there is a derivation in PS of $p'$ from $\Gamma \cup \{p''\}$, then there is a derivation in PS of $p'' \supset p'$ from $\Gamma$ alone.

2. If there is a derivation in PS of $p'$ from $\emptyset \cup \{p''\}$, then there is a derivation in PS of $p'' \supset p'$ from $\emptyset$ [the empty set].

From example 2 it can be seen that the Deduction Theorem implies that if $A \vdash_{PS} B$, then $\vdash_{PS} A \supset B$ (the case where $\Gamma$ is empty).

*Outline of proof.* The idea of the proof is this: We suppose that, for an arbitrary set $\Gamma$ of formulas of P and an arbitrary formula A of P, there is a derivation in PS of a formula B from $\Gamma \cup \{A\}$. We then show how, given such a derivation, to construct a derivation in PS of $A \supset B$ from $\Gamma$ alone. We conclude that *if* there is *any* derivation in PS of B from $\Gamma \cup \{A\}$, *then* there is a derivation in PS of $A \supset B$ from $\Gamma$ alone.

It is essential to the proof that we show how to construct the new derivation given *any* arbitrary derivation of B from $\Gamma \cup \{A\}$. So what we do is, *first*, consider all derivations of B from $\Gamma \cup \{A\}$ that are *exactly one* formula long, and show how to construct, in any such case, a derivation of $A \supset B$ from $\Gamma$ (a derivation that need not be just one formula long). *Then* we show that *if* the Deduction Theorem holds for *all* derivations of B from $\Gamma \cup \{A\}$ that are *less than k* formulas long (where *k* is an arbitrary positive integer), it will also hold for *all* derivations of B from $\Gamma \cup \{A\}$ that are *exactly k* formulas long. And from those two results, and the fact that by definition any derivation in PS is only finitely long, it follows that the Deduction Theorem holds for *all* derivations of B from $\Gamma \cup \{A\}$.

The form of proof described in the last paragraph is known as *proof by* (*strong*) *mathematical induction.* Its two stages are known as the *Basis* and the *Induction Step*. In the Basis of a proof by mathematical induction we establish that the Theorem (whatever it is) holds for the minimal case (in the present proof, the case where the derivation is one formula long: there cannot be a more minimal case than this; a derivation has to be at least one formula long). In the Induction Step we prove that *if* the Theorem holds for all cases up to an arbitrarily given point, *then* it holds also for all cases at the next higher point. (This will be a *strong* induction. A *weak* induction is one in which the Induction Step shows that if the Theorem holds for all cases *at* an arbitrarily given point, then it holds also for all cases at the next higher point.)

*Proof proper*

Let $\Gamma$ be an arbitrary set of formulas of P. Let A be an arbitrary formula of P. Let D be a derivation in PS of a formula B from $\Gamma \cup \{A\}$. D is a finite string of formulas. If there are *n* formulas in D, let them be $D_1, \ldots, D_n$, in the order in which they occur in D; in such a case we shall say that D is of *length n*.

*Basis*: $n = 1$

Let D be a derivation in PS of B from $\Gamma \cup \{A\}$ that is exactly one formula long. We shall show how, given D, we may construct a derivation, which we shall call D′ (and which need not be of length 1), of $A \supset B$ from $\Gamma$ alone.

Since D is a derivation of B and is exactly one formula long, D must in this case consist simply of the formula B itself. That is:

$$D = D_n = D_1 = B.$$

Since there are no formulas in D preceding B, B cannot be an immediate consequence in PS of preceding formulas. So by the definition of *derivation in PS* [§22] B must be either an axiom or in the set $\Gamma \cup \{A\}$. This gives us three cases to consider:

1. B is an axiom.
2. B is in [i.e. is a member of] the set $\Gamma$.
3. B is A itself.

We show, in each of these cases, how to construct a derivation D′ with $A \supset B$ as its last formula and satisfying the condition that it is a derivation from $\Gamma$ alone.

Case 1. B is an axiom. Then D′ is the derivation (of length 3)

| | | |
|---|---|---|
| [1] | B | [Axiom, *ex hypothesi*] |
| [2] | $B \supset (A \supset B)$ | [Axiom, by PS 1] |
| [3] | $A \supset B$ | [MP, 1, 2] |

Here D′ is (not merely a derivation but) a proof in PS, and also therefore a derivation in PS of its last formula from *any* set of formulas of P [cf. 23.5]. [What we have done here is to describe in the metalanguage a sequence of formulas that is a proof in PS. 'A' and 'B' of course are not symbols of P.]

Case 2. B is in the set $\Gamma$. Then D′ is

| | | |
|---|---|---|
| [1] | B | [Given as a member of $\Gamma$] |
| [2] | $B \supset (A \supset B)$ | [Axiom, by PS 1] |
| [3] | $A \supset B$ | [MP, 1, 2] |

Case 3. B is A itself. Then $A \supset B$ is $A \supset A$ and D′ is

| | | |
|---|---|---|
| [1] | $A \supset ((A \supset A) \supset A)$ | [Axiom, by PS 1] |
| [2] | $(A \supset ((A \supset A) \supset A)) \supset ((A \supset (A \supset A)) \supset (A \supset A))$ | [Axiom, by PS 2] |
| [3] | $(A \supset (A \supset A)) \supset (A \supset A)$ | [MP, 1, 2] |
| [4] | $A \supset (A \supset A)$ | [Axiom, by PS 1] |
| [5] | $A \supset A$ | [MP, 4, 3] |

This is a proof of $A \supset A$ in PS, and therefore automatically a derivation in PS of $A \supset A$ from any set of formulas of P you like to take.

*Induction Step*

Assume: The Deduction Theorem holds for every derivation of length less than $k$.

To prove, on this assumption, that: It holds for every derivation of length $k$.

Let D be any derivation of B from $\Gamma \cup \{A\}$ of length $k$. There are four cases to consider:

1. B is an axiom.
2. B is in the set $\Gamma$.
3. B is A itself.
4. B is an immediate consequence by MP of two preceding formulas in D.

Cases 1–3. D′ is exactly as in the Basis.

Case 4. B is an immediate consequence by MP of two preceding formulas in D.

Let these two formulas be $D_i$ and $D_j$, where $i < k$ and $j < k$ (since both $D_i$ and $D_j$ precede B, and B is $D_k$). Then either $D_i$ is $D_j \supset B$, or $D_j$ is $D_i \supset B$, since otherwise B would not be an immediate consequence by MP of $D_i$ and $D_j$. It does not matter which alternative we take, so let us from now on suppose that $D_j$ is $D_i \supset B$.

The length of the derivation of $D_i$ from $\Gamma \cup \{A\}$ is less than $k$. So, by the assumption of the Induction Step, we have

1 $\Gamma \vdash_{PS} A \supset D_i$

[From $\Gamma, A \vdash_{PS} D_i$, by the induction hypothesis]

Similarly we have

2 $\Gamma \vdash_{PS} A \supset D_j$; i.e. $\Gamma \vdash_{PS} A \supset (D_i \supset B)$

But

3 $\vdash_{PS} (A \supset (D_i \supset B)) \supset ((A \supset D_i) \supset (A \supset B))$

[Axiom, by PS 2]

Therefore using 23.4 and 23.5 on 2 and 3

4 $\Gamma \vdash_{PS} (A \supset D_i) \supset (A \supset B)$

and again using 23.4 and 23.5 on 1 and 4

5 $\Gamma \vdash_{PS} A \supset B$

which is what we want.

This completes the Induction Step, and with it the proof of the Deduction Theorem for PS.

[We have shown (in the Basis) that the Deduction Theorem holds for all derivations of length 1, and we have also shown (in the Induction Step) that *if* it holds for all derivations of length less than $k$ (where $k$ is any arbitrary number), *then* it holds for all derivations of length $k$. And this means that we have shown that the Theorem holds for derivations in PS of *any* finite length, i.e. for *all* derivations in PS.]

Analysis of this proof of the Deduction Theorem will show that the only special properties of PS that we appealed to were these:

1. Any wff of the form $A \supset (B \supset A)$ is a theorem.
2. Any wff of the form $(A \supset (B \supset C)) \supset ((A \supset B) \supset (A \supset C))$ is a theorem.
3. Modus Ponens for $\supset$ is the *sole* rule of inference.

So our proof shows that the Deduction Theorem will hold for any other formal system having those three properties (we shall make use of this remark later).

It is clear that if $\vdash_{PS} A \supset B$ then $A \vdash_{PS} B$. From this truth and the Deduction Theorem we get the useful metatheorem:

26.2 $A \vdash_{PS} B$ *iff* $\vdash_{PS} A \supset B$

## 27 Note on proofs by mathematical induction

It is important to get the value of *n* in the Basis right. What we want to prove in an inductive argument is that something holds for *all possible* cases. We prove this by showing (*a*) that it holds for the smallest possible case, and (*b*) that, for any arbitrary case, if it holds for that case, then it holds for the next biggest case. If we are to do this properly our Basis must prove the proposition for the smallest possible case. With the Deduction Theorem the smallest possible case was the case where $n = 1$ [we were concerned with derivations, and a string of formulas of length less than 1 is not a derivation]. But it is not always like this. In later proofs by mathematical induction we shall have Bases where the smallest possible case is the case where $n = 0$. So, for instance, the smallest possible case when we are dealing with the number of connectives in a formula of P is the case where the formula has *no* connectives [in that case it is a single unnegated propositional symbol]. If we take $n = 1$ in the Basis when we ought to have taken $n = 0$, our 'proof' will not prove what we want it to prove: we shall have missed out a possible case.

EXERCISES

1. Give a proof by mathematical induction of the Interpolation Theorem [20.6]:

*If* $\models_P A \supset B$ *and A and B have at least one propositional symbol in common, then there is a formula C of P, all of whose propositional symbols occur in both A and B, such that* $\models_P A \supset C$ *and* $\models_P C \supset B$

[Hint: Use mathematical induction on the number, *n*, of propositional symbols in A but not in B.]

2. Prove by mathematical induction that any four-rowed truth table for any formula A with no connectives other than $\sim$ and $\equiv$ must have in its final column either all T's, or all F's, or two T's and two F's. [Hint: Use mathematical induction on the number, *n*, of occurrences of connectives in A.]

ANSWERS

1. *Proof*

*Basis*: $n = 0$

Then C is A itself.

*Induction Step*

Assume the Theorem holds for all cases where $n < k$. To show it holds for all cases where $n = k$.

Let A be a formula such that $\models_P A \supset B$ and A contains exactly *k* propositional symbols that do not occur in B. Let *p* be one of these symbols. Let *q* be an arbitrary propositional symbol that occurs in both A and B. Let $A_1$ and $A_2$ be the formulas that result from substituting $(q \supset q)$ and $\sim(q \supset q)$ respectively for *p* throughout A. Then $\models_P A_1 \supset B$ and $\models_P A_2 \supset B$. So (by standard truth-table reasoning) $\models_P (\sim A_1 \supset A_2) \supset B$. But $(\sim A_1 \supset A_2) \supset B$ has exactly $k - 1$ propositional symbols in A but not in B. So we can apply the induction hypothesis to it. So there is some formula C, containing only propositional symbols in common to both $(\sim A_1 \supset A_2)$ and B, such that $\models_P (\sim A_1 \supset A_2) \supset C$ and $\models_P C \supset B$. So, given the definitions of $A_1$ and $A_2$, it follows by standard truth-table reasoning that $\models_P A \supset C$. [Think of $\sim A_1 \supset A_2$ as $A_1 \vee A_2$.]

[This last stage of the argument consists in showing that

$$(((\sim A_1 \supset A_2) \supset C) \wedge (C \supset B)) \supset (A \supset C)$$

is a tautology. Standard calculations show that this can only *fail* to be a tautology if A can take the value T while $A_1$ and $A_2$ both take the value F. But from our definitions of $A_1$ and $A_2$, if A has the value T, one or other of $A_1$ and $A_2$ must also have the value T.]

2. *Proof*

*Basis*: $n = 1$ (If $n = 0$ the truth table has only two rows)

Then A has to be of the form B≡C, where B and C are distinct propositional symbols (if they are not distinct, the table has only two rows) (A cannot be of the form ~B where B has no connectives, for then again A would have a table of only two rows). Then the table is

| B | C | B≡C |
|---|---|---|
| T | T | T |
| F | T | F |
| T | F | F |
| F | F | T |

which satisfies the Theorem.

*Induction Step*

Assume the Theorem holds for all formulas with fewer than $k$ occurrences of connectives. To show that it holds for all formulas with exactly $k$ occurrences.

2 cases: 1. A is ~B.
2. A is B≡C.

*Case* 1: A is ~B.
Obvious.

*Case* 2: A is B≡C.
9 subcases, viz.:

1. B has all T's, C has all T's.
2. B has all T's, C has all F's.
3. B has all T's, C has 2 T's and 2 F's.
4. B has all F's, C has all T's.
5. B has all F's, C has all F's.
6. B has all F's, C has 2 T's and 2 F's.
7. B has 2 T's and 2 F's, C has all T's.
8. B has 2 T's and 2 F's, C has all F's.
9. B has 2 T's and 2 F's, C has 2 T's and 2 F's.

Subcases 1–8 are obvious. Subcase 9 can be proved by going through all the finitely many possible combinations.

## 28 Some model-theoretic metatheorems about PS

28.1 *Every axiom of PS is logically valid*

*Proof.* For any interpretation I any formula A is either true or false. So, though the three axiom-schemata of PS represent an infinite set of axioms, the A's, B's and C's in those schemata stand for formulas that can only be either true or false for any given interpretation. So for each axiom-schema we only have to consider a finite number of possibilities; e.g. for schema PS 1 the four possibilities

(1) where A is true and B is true;
(2) where A is false and B is true;
(3) where A is true and B is false;
(4) where A is false and B is false.

In each case $A \supset (B \supset A)$ comes out true: i.e. any formula of P that is an axiom of PS by PS 1 will be true for any interpretation. Similarly for the schemata PS 2 and PS 3.

28.2 *The rule of inference of PS preserves logical validity*
[i.e. anything that is an immediate consequence by it of two formulas both of which are logically valid is itself logically valid]

*Proof.* Directly from 19.6 [if $\models_P A$ and $\models_P A \supset B$, then $\models_P B$].

28.3 *Every theorem of PS is logically valid* [*i.e. if* $\vdash_{PS} A$ *then* $\models_P A$]

*Proof.* Directly from 28.1 and 28.2.

The proof of this metatheorem is in effect a fresh proof of the simple and absolute consistency of PS. Absolute, because there is at least one formula of P that is not logically valid and therefore, by 28.3, not a theorem of PS. Simple, because for no pair of formulas A and $\sim A$ are both A and $\sim A$ logically valid; so, by 28.3, for no pair of formulas A and $\sim A$ are both A and $\sim A$ theorems of PS.

28.4 *If A is a syntactic consequence in PS of a set Γ of formulas of P, then A is a semantic consequence of the set Γ of formulas of P* [*i.e. if* $\Gamma \vdash_{PS} A$, *then* $\Gamma \models_P A$]

*Proof.* Suppose $\Gamma \vdash_{PS} A$. Then by 23.7 there is a finite subset $\Delta$ of $\Gamma$ such that $\Delta \vdash_{PS} A$.

(i) If $\Delta$ is empty, then $\vdash_{PS} A$ and therefore by 28.3 $\vDash_{P} A$, and therefore $\Gamma \vDash_{P} A$ (by 20.5).

(ii) If $\Delta$ is not empty, let $A_1, \ldots, A_n$ $(n \geqslant 1)$ be the members of $\Delta$. Then $A_1, \ldots, A_n \vdash_{PS} A$. Therefore by the Deduction Theorem, applied as many times as may be necessary, $\vdash_{PS} A_1 \supset (A_2 \supset (\ldots (A_n \supset A)))$ [of course if $n = 1$ this will be $\vdash_{PS} A_1 \supset A$]. Therefore by 28.3 $\vDash_{P} A_1 \supset (A_2 \supset (\ldots (A_n \supset A)))$. Therefore, by clause 3 of the definition of *true for an interpretation of P*, there is no model of $\{A_1, \ldots, A_n\}$, i.e. of $\Delta$, that is not also a model of A. Therefore, since $\Delta$ is a subset of $\Gamma$, there is no model of $\Gamma$ that is not also a model of A; i.e. either $\Gamma$ has no model or every model of $\Gamma$ is also a model of A: i.e. $\Gamma \vDash_{P} A$.

For a proof of the converse of 28.4 [i.e. a proof that *if* $\Gamma \vDash_{P} A$, *then* $\Gamma \vdash_{PS} A$], see 32.14 below.

28.5 *Modus Ponens for* $\supset$ *preserves truth-for-an-interpretation I* [i.e. if I is a model of both A and $A \supset B$, then it is a model of B: i.e. Modus Ponens preserves modelhood]

*Proof.* Suppose not. Then, for some interpretation I, A and $A \supset B$ are both true while B is not true. But this is impossible. [Or we could simply havc appealed to 19.5.]

We shall make use of this metatheorem later, especially in §35.

## 29 Concepts of semantic completeness. Importance for logic of a proof of the adequacy and semantic completeness of a formal system of truth-functional propositional logic

*Completeness, in general*

Very roughly, a system is complete if everything you want as a theorem is a theorem. 'As in the case of consistency, the notion of *completeness* of a logistic system has a semantical motivation, consisting roughly in the intention that the system shall have all possible theorems not in conflict with the interpretation. . . . This leads to several purely syntactical definitions of completeness . . .' (Church, 1956, p. 109). We shall put off consideration of syntactic notions of completeness until §33 and concentrate here and in the next few sections on semantic notions of completeness.

What do logicians want? The Holy Grail of logic would be a system or set of systems that caught *all* truths of pure logic. This nobody has yet found. All truths of pure propositional logic then? But they include non-truth-functional logical truths, and the task of systematising these is not yet finished (it is hardly begun). So we ask: 'What do we want that we have some hope of getting?' Here an answer is: 'All truths of pure truth-functional propositional logic'. We shall show that there is a sense in which PS does catch all the truths of pure truth-functional propositional logic, and we now try to make clear what this sense is.

In the first place, the language P of PS is adequate for the expression of any truth function [21.1]. This does not mean that P is capable of expressing any truth about truth functions: it cannot, for instance, express the truth that there are denumerably many distinct truth functions. And it does not mean that every truth-functional tautology (intuitive sense) belongs to P: for P has only the two connectives $\sim$ and $\supset$, and so the tautology $\sim(p' \wedge \sim p')$, for example, does not belong to P. But what we can get from the proofs of metatheorems 21.1 and 21.2 is this: First, any truth-functional formula F with arbitrary connectives could be correlated with a unique formula in disjunctive normal form having the same truth table. [To get uniqueness we should have to add rules, troublesome to state but offering no difficulties in principle, about the order in which various things and steps are to be taken.] Second, any such formula in DNF can be correlated with a unique formula of P having the same truth table. Nothing we say here ensures that there is a 1–1 relation between F and the formula of P: two or more distinct formulas, say F and F′, might both be correlated with the same formula of P. Nevertheless, to any formula F there would be correlated, in what we hope the reader will allow to be a reasonably natural way, a unique formula of P having the same truth table. So (we claim):

Every truth-functional formula is (can be) represented in a natural way by some formula of P.

Second, we shall show (in §31) that every tautology of P is a theorem of PS. Taken in conjunction with what we have just said, this gives us:

Every truth-functional tautology is (can be) represented in a natural way by some theorem of PS.

This is a result that would have delighted the Stoic logicians, who were the first to work systematically on truth-functional propositional logic, and who claimed completeness for their work.[1]

Accordingly we define the following notion of completeness for arbitrary formal systems of truth-functional propositional logic:

*Definition.* A formal system S with language L is *complete with respect to the class of all truth-functional tautologies* iff (1) L is adequate for the expression of any truth function and (2) every tautology of L is a theorem of S.

Similarly for PS:

*Definition.* PS is *complete with respect to the class of all truth-functional tautologies* iff (1) P is adequate for the expression of any truth function and (2) every tautology of P is a theorem of PS.

Usually in the literature proofs of completeness tacitly assume known results about adequacy [for the expression of any truth function] and they concentrate simply on showing that every tautology expressible in the language of the system is a theorem. We shall follow them in this, and so, taking account of the fact that for P the set of tautologies is the same as the set of all logically valid formulas, we frame the following definition:

*Definition.* PS is *semantically complete* iff every logically valid formula of P is a theorem of PS [i.e. iff if $\models_{P} A$ then $\vdash_{PS} A$].

*Remark on terminology*

There is no agreed terminology, and the phrase 'semantic completeness' will not necessarily be understood in the same way by all workers in the field. Most workers simply write

[1] For various reasons it is hard to know whether their claim is correct or not: cf. Benson Mates, *Stoic Logic*, pp. 81–2.

'complete' and then give a brief explanation, or they leave it to the context to reveal what sense of 'complete' they intend.

*On proofs of the semantic completeness of formal systems of truth-functional propositional logic*

All proofs of the *consistency* of formal systems of truth-functional propositional logic are basically the same as Post's. But proofs of their *semantic completeness* are more complicated and more varied. Not counting proofs using topology or algebra,[1] which lie outside the scope of this book, there are at least seven different proofs in the literature: e.g. those of Post (1920), Łukasiewicz (e.g. 1929), László Kalmár (1935), W. V. Quine (1937), Leon Henkin (1947), Kurt Schütte (1954), and Alan Ross Anderson and Nuel Belnap (1959).

## 30 Outline of Post's proof of the semantic completeness of a formal system of truth-functional propositional logic

The first proof of semantic completeness (unless the Stoics got there first: see the last section) was given by Emil Post in 1920 for the propositional system of *Principia Mathematica*. It is simple in principle, but laborious to execute in detail. Post himself, in his 1921 paper, takes only a page and a half over it. But his presentation involves lightning sketches of four mathematical inductions and the acceptance as theorems of about a score of tautologies [Post could take for granted the proofs of these theorems of the system, for they were all in *Principia Mathematica*]. If we were to give proofs in PS of all the corresponding theorems and to present the inductions in the same sort of detail as we did in §26, our proof would be rather long. So, since we shall be giving three different full proofs later, here we shall be content with a bare outline of Post's proof.

*Outline of Post's proof*

For the Whitehead–Russell formulation of the propositional calculus (call it 'PM'):

(1) Post gives an effective method for finding for any formula

[1] e.g. that by Jerzy Łoś (1951). See also the references in fn. 223, p. 285, of Kleene (1967). Stoll (1963, chap. 6, §§8–9) gives a Boolean algebraic treatment of the metatheory of truth-functional propositional logic.

A of PM a formula $A_D$ that is in disjunctive normal form [DNF] and for which $\vdash_{PM} A_D \equiv A$.

(2) Post gives an effective method for proving in PM any tautology of PM that is in DNF.

Now suppose that A is a tautology of PM. If it is already in DNF, then by (2) there is a recipe for proving it in PM. If it is not, then by (1) there is a recipe for finding a formula $A_D$ in DNF such that $\vdash_{PM} A_D \equiv A$, and by (2) there is a recipe for proving $A_D$ in PM. So we have:

$$\vdash_{PM} A_D$$
$$\vdash_{PM} A_D \equiv A$$

From which we get

$$\vdash_{PM} A$$

But A was *any* tautology of PM. So we have:

If A is any tautology of PM, then A is a theorem of PM, i.e. PM is semantically complete.

*Note.* Post's proof not only tells us *that* $\vdash_{PM} A$ (when A is a tautology of PM); it gives us a recipe for constructing a proof in PM of A.

## 31 Proof of the semantic completeness of PS by Kalmár's method

This proof, for a system with additional connectives and some different axioms, was first given by László Kalmár in 1935.

For the proof we need to know that for arbitrary formulas A, B, C of P the following are theorems of PS:

[Item 1] $A \supset A$
[Item 2] $A \supset (B \supset A)$
[Item 3] $(A \supset (B \supset C)) \supset ((A \supset B) \supset (A \supset C))$
[Item 4] $\sim A \supset (A \supset B)$
[Item 5] $A \supset \sim\sim A$
[Item 6] $A \supset (\sim B \supset \sim(A \supset B))$
[Item 7] $(A \supset B) \supset ((\sim A \supset B) \supset B)$

A Kalmár-type proof will work for any system for which all formulas of any of those seven patterns are theorems and MP is a rule of inference. Showing that all formulas of P of any of those seven patterns are theorems is a rather painful exercise involving no new fundamental ideas, and the reader who wants only the basic ideas of the proof should skip the next pages and go straight to the statement of the lemma that is the heart of the proof, i.e. metatheorem 31.14 on p. 100.

Item 1 we have already proved in the course of proving the Deduction Theorem in §26 (see Case 3 of the Basis: p. 86).[1] Items 2 and 3 are just the axiom-schemata PS 1 and PS 2. That leaves Items 4–7 to be proved. It will help in proving these four metatheorems if we first prove the metatheorem 31.1.

The proofs of Items 4–7 are proofs in the metalanguage to the effect that certain formulas are theorems of PS.

Throughout what follows A, B, C are to be arbitrary formulas of P, and to save ink we drop the subscript PS on $\vdash$.

31.1 $A \supset B, B \supset C \vdash A \supset C$

*Proof.* The string consisting of the formulas specified below is a derivation in PS of C from $\{A \supset B, B \supset C, A\}$:

| | | |
|---|---|---|
| 1 | A | [Assumption] |
| 2 | $A \supset B$ | [Assumption] |
| 3 | B | [MP, 1 2] |
| 4 | $B \supset C$ | [Assumption] |
| 5 | C | [MP, 3 4] |

Therefore $A \supset B, B \supset C, A \vdash C$. Therefore by the Deduction Theorem $A \supset B, B \supset C \vdash A \supset C$.

*Abbreviation*
'DT' for 'the Deduction Theorem'.

31.2 [Item 4] $\vdash \sim A \supset (A \supset B)$

| | | | |
|---|---|---|---|
| *Proof.* | 1 | $\vdash \sim A \supset (\sim B \supset \sim A)$ | [PS 1] |
| | 2 | $\vdash (\sim B \supset \sim A) \supset (A \supset B)$ | [PS 3] |
| Therefore | 3 | $\vdash \sim A \supset (A \supset B)$ | [31.1, 1, 2] |

Alternatively, use the formal proof of $\sim p' \supset (p' \supset p'')$ given in example 3 on pp. 73–4, §22, to provide the basis for a proof in the metalanguage that $\vdash \sim A \supset (A \supset B)$

31.3 $\vdash \sim \sim A \supset A$

| | | | |
|---|---|---|---|
| *Proof.* | 1 | $\vdash \sim \sim A \supset (\sim A \supset \sim \sim \sim A)$ | [31.2 = Item 4] |
| | 2 | $\vdash (\sim A \supset \sim \sim \sim A) \supset (\sim \sim A \supset A)$ | [PS 3] |

[1] This amounts to using the *formal* proof of $p' \supset p'$ given in example 1, p. 73, of §22 to provide the basis for a proof *in the metalanguage* that $\vdash_{PS} A \supset A$ for an arbitrary formula A.

| | | |
|---|---|---|
| Therefore 3 | $\vdash \sim\sim A \supset (\sim\sim A \supset A)$ | [31.1, 1, 2] |
| 4 | $\vdash (\sim\sim A \supset (\sim\sim A \supset A)) \supset ((\sim\sim A \supset \sim\sim A) \supset (\sim\sim A \supset A))$ | [PS 2] |
| Therefore 5 | $\vdash (\sim\sim A \supset \sim\sim A) \supset (\sim\sim A \supset A)$ | [MP, 3, 4] |
| 6 | $\vdash \sim\sim A \supset \sim\sim A$ | [Item 1] |
| Therefore 7 | $\vdash \sim\sim A \supset A$ | [MP, 5, 6] |

31.4 [Item 5] $\vdash A \supset \sim\sim A$

| | | |
|---|---|---|
| *Proof.* 1 | $\vdash \sim\sim\sim A \supset \sim A$ | [31.3] |
| 2 | $\vdash (\sim\sim\sim A \supset \sim A) \supset (A \supset \sim\sim A)$ | [PS 3] |
| Therefore 3 | $\vdash A \supset \sim\sim A$ | [MP, 1, 2] |

31.5 $\sim\sim A, A \supset B \vdash \sim\sim B$

*Proof.* $\sim\sim A \vdash \sim\sim A \supset A$, since $\vdash \sim\sim A \supset A$. So there will be a derivation beginning $\sim\sim A$ and ending $\sim\sim A \supset A$, which we represent thus (steps 1 and 2):

| | | |
|---|---|---|
| 1 | $\sim\sim A$ | [Assumption] |
| | . . . | [Gap to be filled by some proof of $\sim\sim A \supset A$, e.g. as indicated in the proof of Metatheorem 31.3] |
| 2 | $\sim\sim A \supset A$ | [31.3] |

Next:

| | | |
|---|---|---|
| 3 | A | [MP, 1, 2] |
| 4 | $A \supset B$ | [Assumption] |
| 5 | B | [MP, 3, 4] |
| | . . . | [Gap to be filled by some proof of $B \supset \sim\sim B$, e.g. as indicated in the proof of 31.4] |
| 6 | $B \supset \sim\sim B$ | [31.4] |
| 7 | $\sim\sim B$ | [MP, 5, 6] |

Therefore $\sim\sim A, A \supset B \vdash \sim\sim B$

31.6 $\vdash (A \supset B) \supset (\sim\sim A \supset \sim\sim B)$

*Proof.* By 31.5 and two applications of the Deduction Theorem.

31.7 $\vdash (A \supset B) \supset (\sim B \supset \sim A)$

| | | |
|---|---|---|
| *Proof.* 1 | $\vdash (A \supset B) \supset (\sim\sim A \supset \sim\sim B)$ | [31.6] |
| 2 | $\vdash (\sim\sim A \supset \sim\sim B) \supset (\sim B \supset \sim A)$ | [PS 3] |
| Therefore 3 | $\vdash (A \supset B) \supset (\sim B \supset \sim A)$ | [31.1, 1, 2] |

31.8 [Item 6] $\vdash A \supset (\sim B \supset \sim(A \supset B))$

*Proof.* By MP A, $A \supset B \vdash B$. Therefore by two applications of DT:

| | | | |
|---|---|---|---|
| | 1 | $\vdash A \supset ((A \supset B) \supset B)$ | |
| | 2 | $\vdash ((A \supset B) \supset B) \supset (\sim B \supset \sim(A \supset B))$ | [31.7] |
| Therefore | 3 | $\vdash A \supset (\sim B \supset \sim(A \supset B))$ | [31.1, 1, 2] |

31.9 $\vdash (\sim A \supset A) \supset (B \supset A)$

| | | | |
|---|---|---|---|
| *Proof.* | 1 | $\vdash \sim A \supset (A \supset \sim B)$ | [Item 4 = 31.2] |
| | 2 | $\vdash (\sim A \supset (A \supset \sim B)) \supset$ $((\sim A \supset A) \supset (\sim A \supset \sim B))$ | [PS 2] |
| Therefore | 3 | $\vdash (\sim A \supset A) \supset (\sim A \supset \sim B)$ | [MP, 1, 2] |
| | 4 | $\vdash (\sim A \supset \sim B) \supset (B \supset A)$ | [PS 3] |
| Therefore | 5 | $\vdash (\sim A \supset A) \supset (B \supset A)$ | [31.1, 3, 4] |

31.10 $\vdash (\sim A \supset A) \supset A$

| | | | |
|---|---|---|---|
| *Proof.* | 1 | $\vdash (\sim A \supset A) \supset ((\sim A \supset A) \supset A)$ | [31.9] |
| | 2 | $\vdash ((\sim A \supset A) \supset ((\sim A \supset A) \supset A)) \supset$ $(((\sim A \supset A) \supset (\sim A \supset A)) \supset ((\sim A \supset A) \supset A))$ | [PS 2] |
| Therefore | 3 | $\vdash ((\sim A \supset A) \supset (\sim A \supset A)) \supset$ $((\sim A \supset A) \supset A)$ | [MP, 1, 2] |
| | 4 | $\vdash (\sim A \supset A) \supset (\sim A \supset A)$ | [Item 1] |
| Therefore | 5 | $\vdash (\sim A \supset A) \supset A$ | [MP, 3, 4] |

31.11 $\vdash (\sim B \supset \sim A) \supset ((\sim A \supset B) \supset (\sim B \supset B))$

*Proof.* The sequence specified below is a derivation in PS of B from $\{\sim B \supset \sim A,\ \sim A \supset B,\ \sim B\}$:

| | | |
|---|---|---|
| 1 | $\sim B$ | [Assumption] |
| 2 | $\sim B \supset \sim A$ | [Assumption] |
| 3 | $\sim A$ | [MP, 1, 2] |
| 4 | $\sim A \supset B$ | [Assumption] |
| 5 | B | [MP, 3, 4] |

Therefore $\sim B \supset \sim A,\ \sim A \supset B,\ \sim B \vdash B$. Therefore by three applications of DT $\vdash (\sim B \supset \sim A) \supset ((\sim A \supset B) \supset (\sim B \supset B))$.

31.12 $A \supset B,\ \sim A \supset B \vdash B$

*Proof.* The string specified below is a derivation of B from $\{A \supset B,\ \sim A \supset B\}$:

| | | |
|---|---|---|
| 1 | $A \supset B$ | [Assumption] |
| ... | [Gap to be filled by a proof of $(A \supset B) \supset (\sim B \supset \sim A)$, as indicated, e.g., in the proof of 31.7] | |
| 2 | $(A \supset B) \supset (\sim B \supset \sim A)$ | [31.7] |

| | | |
|---|---|---|
| 3 | $\sim B \supset \sim A$ | [MP, 1, 2] |
| | ... | [Fill in proof of 31.11] |
| 4 | $(\sim B \supset \sim A) \supset$ $((\sim A \supset B) \supset (\sim B \supset B))$ | [31.11] |
| 5 | $(\sim A \supset B) \supset (\sim B \supset B)$ | [MP, 3, 4] |
| 6 | $\sim A \supset B$ | [Assumption] |
| 7 | $\sim B \supset B$ | [MP, 5, 6] |
| | ... | [Fill in proof of 31.10, with B for A] |
| 8 | $(\sim B \supset B) \supset B$ | [31.10] |
| 9 | B | [MP, 7, 8] |

Therefore $A \supset B, \sim A \supset B \vdash B$.

31.13 [Item 7] $\vdash (A \supset B) \supset ((\sim A \supset B) \supset B)$

*Proof.* From 31.12 by two applications of the Deduction Theorem.

That completes the proof that any formula of the pattern of any of Items 1–7 is a theorem of PS. We can now go on to state and prove the lemma that is the heart of Kalmár's proof. The most difficult thing about the whole proof is understanding exactly what it is that the lemma asserts.

31.14 (*Lemma for the semantic completeness theorem*) *Let A be any formula of P whose only distinct propositional symbols are* $B_1, \ldots, B_k$ $(k \geqslant 1)$. *Let I be an arbitrary interpretation of P. I assigns a set of truth values to the propositional symbols of A, i.e. to* $B_1, \ldots, B_k$. *We define* $B_i^I$ *as follows*:

*If I assigns T to* $B_i$, $B_i^I$ *is to be* $B_i$ *itself*

*If I assigns F to* $B_i$, $B_i^I$ *is to be* $\sim B_i$

*Similarly, let* $A^I$ *be either A or* $\sim A$ *according as A is true or false for I. Then*:

$B_1^I, \ldots, B_k^I \vdash_{PS} A^I$

*Explanation.* Intuitively, what the Lemma says is this:

Let A be any formula of P with *k* distinct propositional symbols. Write out the truth table for A in the standard way. Then for *each row* of the truth table a distinct syntactic consequence relation holds in PS. For *each row* the relation is this: If a propositional symbol is assigned T, then write it to the left of the $\vdash$ sign; if it is assigned F, then write its negation to the left of the $\vdash$ sign. If A is assigned T, write A to the right of the

$\vdash$ sign. If A is assigned F, write $\sim$A to the right of the $\vdash$ sign. *Example.* Let A be $(p' \supset \sim p'') \supset \sim p'''$. It does not matter which of $p'$, $p''$, $p'''$ we take to be $B_1$, $B_2$, $B_3$. So let $B_1 = p'$, $B_2 = p''$, $B_3 = p'''$. Then $A = (B_1 \supset \sim B_2) \supset \sim B_3$ and its truth table is

| $B_1$ | $B_2$ | $B_3$ | A |
|---|---|---|---|
| T | T | T | T |
| F | T | T | F |
| T | F | T | F |
| F | F | T | F |
| T | T | F | T |
| F | T | F | T |
| T | F | F | T |
| F | F | F | T |

Then the Lemma tells us that *eight* syntactic consequence relations hold in the case of A, viz.

[Row 1] $p', p'', p''' \vdash_{PS} (p' \supset \sim p'') \supset \sim p'''$
[Row 2] $\sim p', p'', p''' \vdash_{PS} \sim((p' \supset \sim p'') \supset \sim p''')$
[Row 3] $p', \sim p'', p''' \vdash_{PS} \sim((p' \supset \sim p'') \supset \sim p''')$
[Row 4] $\sim p', \sim p'', p''' \vdash_{PS} \sim((p' \supset \sim p'') \supset \sim p''')$
[Row 5] $p', p'', \sim p''' \vdash_{PS} (p' \supset \sim p'') \supset \sim p'''$
[Row 6] $\sim p', p'', \sim p''' \vdash_{PS} (p' \supset \sim p'') \supset \sim p'''$
[Row 7] $p', \sim p'', \sim p''' \vdash_{PS} (p' \supset \sim p'') \supset \sim p'''$
[Row 8] $\sim p', \sim p'', \sim p''' \vdash_{PS} (p' \supset \sim p'') \supset \sim p'''$

It is vital to see the following point: *Every syntactic consequence relation holds simply in virtue of the syntactic properties of PS.* In *proving* the Lemma we refer to interpretations of P, but every syntactic consequence relation asserted by the Lemma *holds independently of any interpretation of P.* So, e.g., the eight syntactic consequence relations of the example belong to the pure proof theory of PS, and in stating them all references to interpretations of P simply drop out. The presence, both in the statement of the Lemma and in its proof, of references to interpretations of P, tends to obscure this vital point.

*Proof of the Lemma*

The proof is by induction on the number, $n$, of connectives in the formula A. [We drop the subscript on $\vdash$.]

*Basis*: $n = 0$

Then A is a single unnegated propositional symbol, $B_1$. So the Lemma reduces to $B_1 \vdash B_1$ for the case where I assigns T to $B_1$ [i.e. for the row of the truth table for which $B_1$ gets T] and $\sim B_1 \vdash \sim B_1$ for the case where I assigns F to $B_1$. Both these hold by 23.1.

This completes the Basis.

*Induction Step*

Assume that the Lemma holds for all formulas with fewer than $m$ connectives [this is the induction hypothesis]. To prove it holds for any arbitrary formula A with $m$ connectives.

There are two cases to be considered:

1. For some formula C, A is $\sim C$, where C has fewer than $m$ connectives.

2. For some formulas C and D, A is $C \supset D$, where each of C and D has fewer than $m$ connectives.

*Case 1*: A is $\sim C$, where C has fewer than $m$ connectives
2 subcases:

1*a*. A is true for I
1*b*. A is false for I

*Subcase 1a*: A is true for I (so $A^I$ is A)

Then C is false for I (so $C^I$ is $\sim C$).

Since C has fewer than $m$ connectives, we have, by the induction hypothesis

$$B_1^I, \ldots, B_k^I \vdash C^I$$

i.e. $B_1^I, \ldots, B_k^I \vdash \sim C$ (for $C^I$ here is $\sim C$)

i.e. $B_1^I, \ldots, B_k^I \vdash A^I$ (for $A^I$ here $= A = \sim C$)

Which is what we want.

*Subcase 1b*: A is false for I (so $A^I$ is $\sim A$)

Then C is true for I (so $C^I$ is C).

By the induction hypothesis

$$B_1^I, \ldots, B_k^I \vdash C^I$$

i.e. $B_1^I, \ldots, B_k^I \vdash C$ . . . . . 1

But by 31.4 $\vdash C \supset \sim\sim C$ . . . . . 2

Therefore by 1 and 2 and MP

$$B_1^I, \ldots, B_k^I \vdash \sim\sim C$$

i.e. $B_1^I, \ldots, B_k^I \vdash \sim A$

i.e. $B_1^I, \ldots, B_k^I \vdash A^I$

Which is what we want.

*Case 2*: A is $C \supset D$, where each of C and D has fewer than *m* connectives:

3 subcases:

2*a*. C is false for I

2*b*. D is true for I

2c. C is true for I and D is false for I

*Subcase 2a*: C is false for I.

Then A is true for I and $A^I = A = C \supset D$.

By the induction hypothesis

$$B_1^I, \ldots, B_k^I \vdash C^I$$

i.e. $$B_1^I, \ldots, B_k^I \vdash \sim C$$

But by 31.2 $\vdash \sim C \supset (C \supset D)$. So by MP

$$B_1^I, \ldots, B_k^I \vdash C \supset D$$

i.e. $$B_1^I, \ldots, B_k^I \vdash A^I,$$ which is what we want.

*Subcase 2b*: D is true for I.

Then A is true for I and $A^I = A = C \supset D$. By the induction hypothesis

$$B_1^I, \ldots, B_k^I \vdash D$$

But $\vdash D \supset (C \supset D)$ [PS 1]. So by MP

$$B_1^I, \ldots, B_k^I \vdash C \supset D$$

i.e. $$B_1^I, \ldots, B_k^I \vdash A^I,$$ which is what we want.

*Subcase 2c*: C is true for I and D is false for I.

Then A is false for I and $A^I = \sim A = \sim(C \supset D)$. By the induction hypothesis

$$B_1^I, \ldots, B_k^I \vdash C \text{ and also } B_1^I, \ldots, B_k^I \vdash \sim D$$

But $\vdash C \supset (\sim D \supset \sim(C \supset D))$ [31.8]. So by two applications of MP

$$B_1^I, \ldots, B_k^I \vdash \sim(C \supset D)\text{: i.e. } B_1^I, \ldots, B_k^I \vdash A^I.$$

That completes the Induction Step and with it the proof of the Lemma.

31.15 (*The semantic completeness theorem for PS*) *Every logically valid formula of P is a theorem of PS* [*or*: *If* $\vDash_P A$, *then* $\vdash_{PS} A$]

*Proof*. Let A be an arbitrary logically valid formula of P with distinct propositional symbols $B_1, \ldots, B_k$ ($k \geqslant 1$).

Let I and J be interpretations of P that differ only in that I assigns T to $B_k$ while J assigns F to $B_k$. Then by the Lemma

(1) $B_1^I, \ldots, B_k^I \vdash A^I$
and
(2) $B_1^J, \ldots, B_k^J \vdash A^J$.
Now we have from our definitions of I, J, etc.,
(3) $B_1^J = B_1^I, \ldots, B_{k-1}^J = B_{k-1}^I$
and
(4) $B_k^I = B_k$
and
(5) $B_k^J = \sim B_k$.
Also, since A is logically valid it is true for I and true for J. So
(6) $A^I = A^J = A$.
From (1) and (4) and (6) we get
(7) $B_1^I, \ldots, B_{k-1}^I, B_k \vdash A$.
From (2) and (3) and (5) and (6) we get
(8) $B_1^I, \ldots, B_{k-1}^I, \sim B_k \vdash A$.
By applying the Deduction Theorem to (7) and (8) we get
(9) $B_1^I, \ldots, B_{k-1}^I \vdash B_k \supset A$
and
(10) $B_1^I, \ldots, B_{k-1}^I \vdash \sim B_k \supset A$.
But we have
(11) $\vdash (B_k \supset A) \supset ((\sim B_k \supset A) \supset A)$ [31.13 = Item 7]
So by two applications of MP we get
(12) $B_1^I, \ldots, B_{k-1}^I \vdash A$.

We have now eliminated $B_k$ from the argument. If $k = 1$, we have immediately $\vdash_{PS} A$. If $k > 1$, then let L be an interpretation that differs from I only in the truth value it assigns to $B_{k-1}$ (i.e. if I assigns T to $B_{k-1}$, then L assigns F to $B_{k-1}$; and if I assigns F to $B_{k-1}$, then L assigns T to $B_{k-1}$). By repeating the set of moves made above, with obvious changes, we can eliminate $B_{k-1}$ from the argument just as we eliminated $B_k$. And so on, until we have eliminated everything to the left of the '$\vdash$' sign and are left simply with

$$\vdash_{PS} A.$$

But A was an arbitrary logically valid formula of P. So if $\vDash_P A$, then $\vdash_{PS} A$. Q.E.D.

Analysis of this proof of completeness shows that *any* formal system with the language P that has Modus Ponens as a rule of inference and satisfies the following seven conditions [Items 1–7] will be semantically complete: For arbitrary formulas A, B, C:

1. $\vdash A \supset A$
2. $\vdash A \supset (B \supset A)$
3. $\vdash (A \supset (B \supset C)) \supset ((A \supset B) \supset (A \supset C))$
4. $\vdash \sim A \supset (A \supset B)$
5. $\vdash A \supset \sim\sim A$
6. $\vdash A \supset (\sim B \supset \sim(A \supset B))$
7. $\vdash (A \supset B) \supset ((\sim A \supset B) \supset B)$

## 32 Proof of the semantic completeness of PS by Henkin's method

In this section we prove the semantic completeness of PS by a different method. We do this for several reasons: (1) The method is essentially the same as, but simpler than, the method used in proving the semantic completeness of the system of predicate logic in Part 3, and familiarity with it will make the general pattern of the very complicated later proof easier to grasp. (2) The proof makes use of a new notion, that of a maximal p-consistent set, that is of general utility. (3) In the course of the proof we establish two very useful metatheorems, one of them Lindenbaum's Lemma for PS, the other the theorem that every p-consistent set of PS has a model [32.13]. The second enables us to make a complete tie-up between the proof theory of PS and the model theory of P, and to prove among other things a 'strong' completeness theorem for PS. (4) Finally, since some parts of the proof have exact parallels in the later proof of the semantic completeness of our system of predicate logic, we can shorten the statement of that later proof by citing the appropriate parts of this one, without going through all the details again.

The proof, for a system of predicate logic, was first given by Leon Henkin in 1947 [Henkin, 1947]. Subsections (*a*), (*b*), (*c*) and (*d*) establish essential preliminaries to the main part of the proof, which comes in subsection (*e*).

### (*a*) *Proof that certain formulas are theorems of PS*

In the proof we use the fact that for any arbitrary formulas A, B, C of P the following are theorems of PS:

1. $A \supset A$
2. $A \supset (B \supset A)$

3. $(A \supset (B \supset C)) \supset ((A \supset B) \supset (A \supset C))$
4. $\sim A \supset (A \supset B)$
5. $A \supset (\sim B \supset \sim (A \supset B))$
6. $(A \supset B) \supset ((A \supset \sim B) \supset \sim A)$
7. $(\sim A \supset B) \supset ((\sim A \supset \sim B) \supset A)$

The first five of these we have proved already [as Items 1, 2, 3, 4 and 6 in §31]. We now show that 6 and 7 also hold [Metatheorems 32.4 and 32.5 below]. We begin by stating a metatheorem that can be seen to be true in virtue of the definition of $\vdash_{PS}$ (in fact it holds for arbitrary formal systems).

32.1 *If* $A, B \vdash_{PS} C$ *and* $A, C \vdash_{PS} D$, *then* $A, B \vdash_{PS} D$

[Subscript on $\vdash$ dropped in what follows]

32.2 $\vdash (A \supset \sim B) \supset (B \supset \sim A)$

*Proof.* 1 $\sim\sim A \supset A, A \supset \sim B \vdash \sim\sim A \supset \sim B$ [31.1]
2 $\vdash (\sim\sim A \supset \sim B) \supset (B \supset \sim A)$ [PS 3]
Therefore 3 $\sim\sim A \supset A, A \supset \sim B \vdash B \supset \sim A$ [MP, 1, 2]
Therefore 4 $\vdash (\sim\sim A \supset A) \supset ((A \supset \sim B) \supset (B \supset \sim A))$ [3, DT twice]
5 $\vdash \sim\sim A \supset A$ [31.3]
Therefore 6 $\vdash (A \supset \sim B) \supset (B \supset \sim A)$ [MP, 5, 4]

32.3 $\vdash (A \supset \sim A) \supset \sim A$

*Proof.* 1 $\vdash (A \supset \sim A) \supset (\sim\sim A \supset \sim A)$ [31.7]
2 $\vdash (\sim\sim A \supset \sim A) \supset \sim A$ [31.10]
3 $(A \supset \sim A) \supset (\sim\sim A \supset \sim A), (\sim\sim A \supset \sim A) \supset \sim A \vdash (A \supset \sim A) \supset \sim A$ [31.1]
Therefore 4 $\vdash ((A \supset \sim A) \supset (\sim\sim A \supset \sim A)) \supset (((\sim\sim A \supset \sim A) \supset \sim A) \supset ((A \supset \sim A) \supset \sim A))$ [3, DT twice]
Therefore 5 $\vdash ((\sim\sim A \supset \sim A) \supset \sim A) \supset ((A \supset \sim A) \supset \sim A)$ [MP, 1, 4]
Therefore 6 $\vdash (A \supset \sim A) \supset \sim A$ [MP, 2, 5]

32.4 $\vdash (A \supset B) \supset ((A \supset \sim B) \supset \sim A)$

*Proof.* 1 $A \supset B, B \supset \sim A \vdash A \supset \sim A$ [31.1]
2 $\vdash (A \supset \sim A) \supset \sim A$ [32.3]

Therefore 3 $A \supset B, B \supset \sim A \vdash \sim A$ [MP, 1, 2]

4 $A \supset B, A \supset \sim B \vdash A \supset \sim B$ [General property of $\vdash$: cf. 23.1 and 23.2]

5 $\vdash (A \supset \sim B) \supset (B \supset \sim A)$ [32.2]

Therefore 6 $A \supset B, A \supset \sim B \vdash B \supset \sim A$ [MP, 4, 5]

Therefore 7 $A \supset B, A \supset \sim B \vdash \sim A$ [32.1, 6, 3]

Therefore 8 $\vdash (A \supset B) \supset ((A \supset \sim B) \supset \sim A)$ [7, DT twice]

32.5 $\vdash (\sim A \supset B) \supset ((\sim A \supset \sim B) \supset A)$

*Proof.* 1 $\sim A \supset B, \sim A \supset \sim B \vdash \sim \sim A$ [Step 7 of proof of 32.4, with $\sim A$ for A]

2 $\vdash \sim \sim A \supset A$ [31.3]

Therefore 3 $\sim A \supset B, \sim A \supset \sim B \vdash A$ [MP, 1, 2]

Therefore 4 $\vdash (\sim A \supset B) \supset ((\sim A \supset \sim B) \supset A$ [3, DT twice]

(*b*) *p-consistent sets, maximal p-consistent sets, and some theorems about them*

We remind the reader that a set $\Gamma$ of formulas of P is a p-consistent set of PS iff for no formula A of P is it the case that both $\Gamma \vdash_{PS} A$ and $\Gamma \vdash_{PS} \sim A$.

32.6 *If a set of formulas of P has a model, then it is a p-consistent set of PS*

*Proof.* Let $\Gamma$ be any set of formulas of P. Suppose $\Gamma$ has a model but is not a p-consistent set of PS. Then for some formula A of P $\Gamma \vdash_{PS} A$ and $\Gamma \vdash_{PS} \sim A$. Then by 28.4 [if $\Gamma \vdash_{PS} A$, then $\Gamma \vDash_P A$] $\Gamma \vDash_P A$ and $\Gamma \vDash_P \sim A$; i.e. every model of $\Gamma$ is a model of A and every model of $\Gamma$ is a model of $\sim A$. But $\Gamma$ has a model, by assumption. So in that model A is true and $\sim A$ is true. But this is impossible. Therefore if $\Gamma$ has a model it is a p-consistent set of PS.

The converse of 32.6 is important and beautiful but much harder to prove. It is the crucial lemma for any Henkin-type completeness proof, and it takes us up to 32.13 to prove it.

32.7 *$\Gamma \cup \{\sim A\}$ is a p-inconsistent set of PS iff $\Gamma \vdash_{PS} A$*

*Proof*

(*a*) Suppose $\Gamma \vdash_{PS} A$. Then $\Gamma, \sim A \vdash_{PS} A$ [23.2]. But $\Gamma, \sim A \vdash_{PS} \sim A$ [23.1, 23.2]. So $\Gamma \cup \{\sim A\}$ is a p-inconsistent set of PS.

(*b*) Suppose $\Gamma \cup \{\sim A\}$ is a p-inconsistent set of PS. Then $\Gamma, \sim A \vdash_{PS} B$ and $\Gamma, \sim A \vdash_{PS} \sim B$ for some formula B. Therefore, by the Deduction Theorem, $\Gamma \vdash_{PS} \sim A \supset B$ and $\Gamma \vdash_{PS} \sim A \supset \sim B$. But $\vdash_{PS} (\sim A \supset B) \supset ((\sim A \supset \sim B) \supset A)$ [32.5]. So, by Modus Ponens twice, $\Gamma \vdash_{PS} A$.

32.8 *$\Gamma \cup \{A\}$ is a p-inconsistent set of PS iff $\Gamma \vdash_{PS} \sim A$*

*Proof.* As for 32.7, but using 32.4 [$\vdash_{PS} (A \supset B) \supset ((A \supset \sim B) \supset \sim A)$] in (*b*).

*Definition.* $\Gamma$ is a *maximal p-consistent set of PS* iff $\Gamma$ is a p-consistent set of PS and if A is any arbitrary formula of P, then either A is a member of $\Gamma$ or $\Gamma, A \vdash_{PS} B$ and $\Gamma, A \vdash_{PS} \sim B$ for some formula B of P.

Informally, a maximal p-consistent set of PS is a p-consistent set of PS to which no formula can be added without p-inconsistency: it has got all the formulas it can take.

32.9 *For any maximal p-consistent set $\Gamma$ of PS and any formula A of P, exactly one of A and $\sim A$ is in $\Gamma$*

*Proof.* Obviously they cannot both be in $\Gamma$, since $\Gamma$ is a p-consistent set of PS. Suppose neither is. Then neither can be added to $\Gamma$ without p-inconsistency; i.e. both $\Gamma \cup \{A\}$ and $\Gamma \cup \{\sim A\}$ are p-inconsistent sets of PS. Therefore by 32.8 $\Gamma \vdash_{PS} \sim A$ and by 32.7 $\Gamma \vdash_{PS} A$; i.e. $\Gamma$ is a p-inconsistent set of PS. But $\Gamma$ was a p-consistent set *ex hypothesi*. So exactly one of A and $\sim A$ is in $\Gamma$.

32.10 *For any maximal p-consistent set $\Gamma$ of PS and any formula A of P, if $\Gamma \vdash_{PS} A$ then A is a member of $\Gamma$*

*Proof.* Suppose $\Gamma \vdash_{PS} A$ but that A is not a member of $\Gamma$. Then by 32.9 $\sim A$ is a member of $\Gamma$. Therefore $\Gamma \vdash_{PS} \sim A$. Therefore $\Gamma$ is a p-inconsistent set of PS. But $\Gamma$ was a p-consistent set *ex hypothesi*. So if $\Gamma \vdash_{PS} A$ then A is a member of $\Gamma$.

### (*c*) *Enumeration Theorem for P*

32.11 (*Enumeration Theorem*) *The formulas of P are effectively enumerable*

*Proof.* Assign to the symbol $p$ the numeral 10
" " " " $'$ " " 100

| | | | |
|---|---|---|---|
| Assign to the symbol | $\sim$ | the numeral | 1000 |
| ,, ,, ,, ,, | $\supset$ | ,, ,, | 10000 |
| ,, ,, ,, ,, | ( | ,, ,, | 100000 |
| ,, ,, ,, ,, | ) | ,, ,, | 1000000 |

[It does not matter from the theoretical point of view whether these numerals are binary or decimal, provided we keep to the same system throughout.]

Let the number of a *formula* be the number denoted by the numeral arrived at by juxtaposing in order from left to right the numerals of the symbols in the formula, as in the example that follows.

*Example.* The number of the formula $\sim(p' \supset p'')$ is the number denoted by the numeral

$$10001000001010010000101001001000000$$

Each distinct formula of P will thus have a distinct positive integer assigned to it, and we order the formulas in the order of their associated numbers. This gives us an enumeration of the formulas.

We show next that this enumeration is *effective*:

*There is an effective method for finding, for any positive integer n, the nth formula in the enumeration*

*Proof.* The smallest number assigned to a formula of P is the number 10100, which is assigned to the formula $p'$. Write down $p'$ as the first formula in the enumeration. Then take the numbers from 10100 on in succession, checking each to see if it is the number of a formula [the definition of formula in §17 can be turned into an effective method for discriminating formulas from non-formulas]. If it is, write down that formula in your list, keeping a check on how many formulas you have written down. If it is not, go on to the next number. The *n*th formula you write down is the one you want.

Further: Given any formula there is an effective method for finding for which number *n* it is the *n*th formula in the enumeration. (*Proof.* Find the number of the formula. Go through all the numbers from 10100 on up to that number, listing any formulas they are the numbers of, as in the proof above. The number of formulas in the list up to and including the given formula is the number you want.)

### (*d*) *Lindenbaum's Lemma for PS*

32.12 (*Lindenbaum's Lemma for PS*) *Any p-consistent set of PS is a subset of some maximal p-consistent set of PS*

['some': i.e. some or other]

*Historical note*. The corresponding theorem for a system of predicate logic was first proved by Adolf Lindenbaum, a Polish mathematician and logician, who was killed by the Nazis in the summer of 1941 [cf. Tarski, 1923–38, p. 98, Theorem 56].

*Proof*. Let $\Gamma$ be any p-consistent set of PS. We define an infinite sequence of sets $\langle \Gamma_0, \Gamma_1, \Gamma_2, \ldots \rangle$ as follows:

Let $\Gamma_0$ be $\Gamma$. Let $A_{n+1}$ be the $(n+1)$th formula in our enumeration [32.11]. Let $\Gamma_{n+1}$ be $\Gamma_n \cup \{A_{n+1}\}$ if this is a p-consistent set of PS; otherwise let $\Gamma_{n+1}$ be $\Gamma_n$.

[Informally: If the first formula in our enumeration can consistently be added to $\Gamma$, then add it to $\Gamma$ to make $\Gamma_1$; if it cannot, then $\Gamma_1$ is to be $\Gamma$ itself. If the second formula in our enumeration can consistently be added to $\Gamma_1$, then add it to $\Gamma_1$ to make $\Gamma_2$; if it cannot, then $\Gamma_2$ is to be $\Gamma_1$. And so on.]

We now prove that the set that is the union of all the sets $\Gamma_0, \Gamma_1, \Gamma_2, \ldots$ [i.e. the set $\Gamma_0 \cup \Gamma_1 \cup \Gamma_2 \cup \ldots$] is a maximal p-consistent set of PS. Let this set be $\Gamma'$. Obviously $\Gamma$ is a subset of $\Gamma'$. We prove first that $\Gamma'$ is a p-consistent set of PS; then that it is a maximal p-consistent set of PS.

(i) *$\Gamma'$ is a p-consistent set of PS*

*Proof*

(*a*) Each of the $\Gamma_i$'s is p-consistent; for $\Gamma_0$ is p-consistent (*ex hypothesi*) and, by the way we constructed the $\Gamma_i$'s, if $\Gamma_n$ is p-consistent then so is $\Gamma_{n+1}$.

(*b*) Now suppose $\Gamma'$ is p-inconsistent. Then for some formula A there is a derivation of A from $\Gamma'$ and also a derivation of $\sim A$ from $\Gamma'$. Each of those two derivations consists of a finite number of formulas. Let $A_n$ be the highest numbered formula in our enumeration [32.11] that occurs in either of the derivations, and let it be the *n*th formula in the enumeration. Then $\Gamma_n \vdash_{PS} A$ and $\Gamma_n \vdash_{PS} \sim A$; i.e. $\Gamma_n$ is p-inconsistent. But this contradicts (*a*) above. So $\Gamma'$ must be p-consistent.

*Comment*. This simple fact, that every proof and every derivation in the formal systems we are concerned with must consist of only a *finite* number of formulas, is perhaps second only to mathematical induction in its usefulness for metatheory [we have already used it in the proof of 28.4].

(ii) *Γ′ is a maximal p-consistent set of PS*

*Proof.* Let $A_n$ be any formula of P, with *n* the number of its place in our enumeration. Suppose $A_n$ is not a member of Γ′. Then $\Gamma_{n-1} \cup \{A_n\}$ must be a p-inconsistent set, since otherwise $A_n$ would have been added to $\Gamma_{n-1}$ to make $\Gamma_n$. So $\Gamma_{n-1}, A_n \vdash_{PS} B$ and $\Gamma_{n-1}, A_n \vdash_{PS} {\sim}B$ for some formula B. Therefore $\Gamma', A_n \vdash_{PS} B$ and $\Gamma', A_n \vdash_{PS} {\sim}B$ for some B (since $\Gamma_{n-1}$ is a subset of Γ′). Therefore either $A_n$ is in Γ′ or both $\Gamma', A_n \vdash_{PS} B$ and $\Gamma', A_n \vdash_{PS} {\sim}B$ for some B. That is, Γ′ is a maximal p-consistent set of PS.

That completes the proof of Lindenbaum's Lemma, and with it the preliminaries to the main part of the proof.

(*e*) *The main part of the proof*

32.13 *Every p-consistent set of PS has a model*

*Proof.* Let Γ be any p-consistent set of PS. By Lindenbaum's Lemma there is some maximal p-consistent set, Γ′, that has Γ as a subset. If Γ′ has a model then so does Γ, since every formula in Γ is also in Γ′. We show that Γ′ has a model. Two stages: (1) We give an interpretation I for P. (2) We show that, for any arbitrary formula A, if A is in Γ′ then A is true for the interpretation I. [In fact we prove something stronger: we prove that A is true for I *iff* A is in Γ′. We prove this stronger proposition not just for the fun of it, but because we need an 'iff' in the induction hypothesis in the proof of the weaker one. We have to prove the strong proposition in order to get the weak one.]

*Stage 1*: *Interpretation for P*

Our interpretation I assigns T to every propositional symbol in [i.e. that is a member of] Γ′, F to every other propositional symbol. The connectives have their standard meanings.

*Stage 2*: *A is true for I iff A is in* Γ′

*Proof.* The proof is by induction on the number, *n*, of connectives in A.

*Basis*: $n = 0$

Then A is a propositional symbol, and so, by our assignment, A is true for I iff A is in $\Gamma'$.

*Induction Step*

Assume that the theorem [viz. A is true for I iff A is in $\Gamma'$] holds for every formula with fewer than $m$ connectives [induction hypothesis]. To prove it holds for every formula with $m$ connectives.

2 cases:

1. A is $\sim$B, where B has fewer than $m$ connectives.
2. A is $B \supset C$, where each of B and C has fewer than $m$ connectives.

In each case we want to prove [1st leg] if A is true for I then A is in $\Gamma'$, and [2nd leg] if A is in $\Gamma'$ then A is true for I.

*Case 1*: *A is* $\sim B$

1st leg: Assume A is true for I. Then B is false for I. So by the induction hypothesis B is not in $\Gamma'$. Therefore, by 32.9, $\sim$B is in $\Gamma'$; i.e. A is in $\Gamma'$.

2nd leg: Assume A is in $\Gamma'$. Then B is not in $\Gamma'$. Therefore by the induction hypothesis B is not true for I. Therefore A is true for I.

*Case 2*: A is $B \supset C$

1st leg: Assume A is true for I. Then either B is not true for I or C is true for I. Therefore either B is not in $\Gamma'$ or C is in $\Gamma'$.

(i) Suppose B is not in $\Gamma'$. Then $\sim$B is in $\Gamma'$. So $\Gamma' \vdash_{PS} \sim B$. Then by 31.2 [$\vdash_{PS} \sim B \supset (B \supset C)$] and MP $\Gamma' \vdash_{PS} B \supset C$. So by 32.10 $B \supset C$ is in $\Gamma'$; i.e. A is in $\Gamma'$.

(ii) Suppose, alternatively, C is in $\Gamma'$. Then $\Gamma' \vdash_{PS} C$. Then by PS 1 [$\vdash_{PS} C \supset (B \supset C)$] and MP $\Gamma' \vdash_{PS} B \supset C$ and so $B \supset C$ is in $\Gamma'$; i.e. A is in $\Gamma'$.

2nd leg: Assume A is not true for I. Then B is true for I and C is false for I. Hence B is in $\Gamma'$ and C is not in $\Gamma'$. So $\sim$C is in $\Gamma'$. So $\Gamma' \vdash_{PS} B$ and $\Gamma' \vdash_{PS} \sim C$. Hence, by 31.8 [$\vdash_{PS} B \supset (\sim C \supset \sim(B \supset C))$] and MP twice, $\Gamma' \vdash_{PS} \sim(B \supset C)$. So $\sim(B \supset C)$ is in $\Gamma'$; i.e. $\sim$A is in $\Gamma'$. So A is not in $\Gamma'$. So: If A is in $\Gamma'$, then A is true for I.

That completes the Induction Step and the proof of 32.13.

32.14 (*The 'strong' completeness theorem for PS*) *If* $\Gamma \vDash_P A$ *then* $\Gamma \vdash_{PS} A$

*Proof.* Suppose $\Gamma \vDash_P A$. Then $\Gamma \cup \{\sim A\}$ has no model. Therefore, by 32.13, $\Gamma \cup \{\sim A\}$ is not a p-consistent set of PS, and obviously therefore it is a p-inconsistent set of PS. So, by 32.7, $\Gamma \vdash_{PS} A$.

32.15 (*The semantic completeness theorem for PS*) *If* $\vDash_P A$ *then* $\vdash_{PS} A$

*Proof.* In 32.14 take the empty set for $\Gamma$.

32.13 and 32.14 (the strong completeness theorem) do not come out of the Kalmár-type proof of completeness.

The power of 32.13 is still further revealed in Metatheorems 32.16–32.21, all proved with its help.

32.16 *A set of formulas of P is a p-consistent set of PS iff it has a model*

*Proof.* Directly from 32.13 and 32.6.

32.17 $\Gamma \vDash_P A$ *iff* $\Gamma \vdash_{PS} A$

*Proof.* Directly from 32.14 and 28.4 [if $\Gamma \vdash_{PS} A$ then $\Gamma \vDash_{PS} A$]

This last result means that we can freely interchange $\vDash_P$ and $\vdash_{PS}$ in any result we already have. So:

32.18 (*The Finiteness Theorem for P*) $\Gamma \vDash_P A$ *iff there is a finite subset* $\Delta$ *of* $\Gamma$ *such that* $\Delta \vDash_P A$

*Proof.* From 23.7 [The analogue for $\vdash_{PS}$ of 32.18] and 32.17.

This is rather a surprising result, considering that $\Gamma$ may be an infinite set. It is obvious in one direction [viz. if $\Delta$ is a finite subset of $\Gamma$ then if $\Delta \vDash_P A$ then $\Gamma \vDash_P A$]. It is the other direction [if $\Gamma \vDash_P A$ then there is a finite subset $\Delta$ of $\Gamma$ such that $\Delta \vDash_P A$] that is unexpected.

32.19 *If every finite subset of a set* $\Gamma$ *of formulas of P is a p-consistent set of PS, then* $\Gamma$ *is a p-consistent set of PS*

*Proof.* Suppose the antecedent is true but that $\Gamma$ is not a p-consistent set of PS (and so is p-inconsistent). Then for some formula A $\Gamma \vdash_{PS} A$ and $\Gamma \vdash_{PS} \sim A$. So there is some derivation in PS of A from $\Gamma$ and some derivation in PS of $\sim A$ from $\Gamma$. By our definition of *derivation in PS*, each of these derivations consists of only finitely many formulas, and so only finitely many formulas of Γ occur in them. So there is a finite subset Δ of Γ such that $\Delta \vdash_{PS} A$ and $\Delta \vdash_{PS} \sim A$. But this contradicts our hypothesis that the antecedent was true.

From 32.19 and 32.16 there follows the result known as the Compactness Theorem:

32.20 (*The Compactness Theorem for P*) *If every finite subset of a set Γ of formulas of P has a model, then Γ has a model*

*Proof.* 32.16 and 32.19.

Since the converse of 32.20 is obviously true, we could write 32.20 with an 'iff' instead of an 'if'.

32.17 applied to 20.6 yields the syntactic version of the Interpolation Theorem:

32.21 (*The Interpolation Theorem for PS*) *If* $\vdash_{PS} A \supset B$, *and A and B have at least one propositional symbol in common, then there is a formula C of P, all of whose propositional symbols occur in both A and B, such that* $\vdash_{PS} A \supset C$ *and* $\vdash_{PS} C \supset B$

*Proof.* From 20.6 and 32.17.

In short, what we have is an exact correspondence between the proof theory of PS and the model theory of P. The key theorem for proving this correspondence is 32.13: *Every p-consistent set of PS has a model.*

The Compactness Theorem, 32.20, is an interesting result, and in its version for predicate logic it probably has even more applications than the corresponding completeness theorem; for instance, we use it in §48 to prove the existence of a non-standard model. The statement of the theorem at 32.20 makes no reference to any deductive apparatus, and the theorem can be given a purely model-theoretic proof. The same is true for the version for predicate logic as originally formulated by Gödel (1930, Theorem X); our own formulation at 45.20 is less pure in this respect. For interest, we give a purely model-theoretic proof of the propositional version.

*Model-theoretic proof of the Compactness Theorem for P* (32.20)

We define a *partial interpretation* of P as an assignment of truth values to at most a proper subset of the propositional symbols of P. We say that an interpretation or partial interpretation $X'$ of P is an *extension* of an interpretation or partial interpretation $X$ of P iff each assignment of a truth value to a propositional symbol that $X$ makes is also made by $X'$. We say that a partial interpretation $X$ *falsifies* a set $\Delta$ of formulas of P iff there is no extension of $X$ that is a model of $\Delta$. We write $p_1$ for $p'$, $p_2$ for $p''$, etc.

Now suppose that every finite subset of a set $\Gamma$ of formulas of P has a model. We define a sequence of partial interpretations, $X_0, X_1, X_2, \ldots$, as follows:

$X_0$ is an assignment of truth values simply to the empty set, i.e. it does not assign any truth value to any propositional symbol of P.

$X_{i+1}$ is to be $X_i$ plus the assignment of T to $p_{i+1}$ if the resulting partial interpretation does not falsify any finite subset of $\Gamma$; otherwise, $X_{i+1}$ is to be $X_i$ plus the assignment of F to $p_{i+1}$.

We show by induction that none of the $X_i$'s falsifies any finite subset of $\Gamma$. Basis: $X_0$ does not falsify any finite subset of $\Gamma$. Induction hypothesis: $X_i$ does not falsify any finite subset of $\Gamma$. To prove that $X_{i+1}$ also does not falsify any. Suppose $X_{i+1}$ falsifies some finite subset $\Delta$ of $\Gamma$. Then, by construction, $X_{i+1}$ must be $X_i$ plus the assignment of F to $p_{i+1}$, and furthermore $X_i$ plus the assignment of T to $p_{i+1}$ must falsify some finite subset, say $\Sigma$, of $\Gamma$. So we have:

No extension of $X_i$ together with the assignment of F to $p_{i+1}$ is a model of $\Delta$, or consequently of $\Delta \cup \Sigma$.

No extension of $X_i$ together with the assignment of T to $p_{i+1}$ is a model of $\Sigma$, or consequently of $\Delta \cup \Sigma$.

So no extension of $X_i$ is a model of $\Delta \cup \Sigma$. But $\Delta \cup \Sigma$ is a finite subset of $\Gamma$. So $X_i$ falsifies a finite subset of $\Gamma$. This contradicts the induction hypothesis. So our supposition that $X_{i+1}$ falsified some finite subset of $\Gamma$ must be wrong, which is what we wanted to prove.

Let $X$ be the interpretation of P got by taking all the $X_i$'s together. Suppose $\Gamma$ has no model. Then some formula A in $\Gamma$ is false for $X$. Let $p_k$ be the highest-numbered propositional symbol (i.e. the one with the most dashes) occurring in A. Then $X_k$ falsifies {A}, a finite subset of $\Gamma$. This contradicts the result obtained in the previous paragraph, that none of the $X_i$'s falsifies any finite subset of Γ. So Γ has a model.

Q.E.D.

## 33 Concepts of syntactic completeness. Proof of the syntactic completeness (in one sense) of PS

There are several concepts of syntactic completeness. One natural one is the following:

A formal system S is complete iff for each formula A (of the language of the system) either A or $\sim$A is a theorem of S.[1]

PS is not complete in that sense. For example, neither $p'$ nor $\sim p'$ is a theorem of PS (only tautologies are theorems of PS, and neither $p'$ nor $\sim p'$ is a tautology). It would be easy enough to construct a formal system with language P that was complete in that sense: e.g. the system that has for its axioms every (unnegated) propositional symbol as well as the axioms of PS. But the logician wants as theorems only formulas that are logically valid, and in the system just described many of the theorems would not be logically valid. From the logician's point of view PS is perfectly all right as it is: it has as theorems all logically valid formulas of P and only logically valid formulas of P, and it is adequate for the expression of any truth function.

However, there is a sense of 'syntactically complete' in which PS is syntactically complete:

*Definition.* PS is *syntactically complete* (in one sense) iff no unprovable schema can be added to it as an axiom-schema without inconsistency.

Here and below we assume as obvious the notions of *schema*, *provable schema*, *adding as an axiom-schema*, *tautological schema*, etc.

33.1 *PS is syntactically complete*

*Proof.* Let U be any schema unprovable in PS, with an arbi-

[1] We make frequent use in Parts 3 and 4 of a modification of this concept that we call *negation-completeness*: see e.g. the definition before 45.10 and metatheorems 45.10, 45.13, 45.14, 46.4, 48.3, 51.12, 51.13 (The Generalised Gödel Theorem).

trary number, $k$, of distinct schematic letters, $U_1, \ldots, U_k$. Then, by the semantic completeness theorem, U cannot be a tautological schema (if it were it would be provable). So there is some assignment (call it 'V') of truth values to the schematic letters of U for which, on the standard truth-table evaluation, U as a whole gets the value F. Let U be added to PS as an axiom-schema, and let PS* be this enlargement of PS. Then every substitution of formulas $A_1, \ldots, A_k$ (not necessarily distinct) for the schematic letters $U_1, \ldots, U_k$ of U will be an axiom, and therefore a theorem, of PS*. Let B be the formula that results from substituting $p' \supset p'$ or $\sim(p' \supset p')$ for each $U_i$ in U according as the assignment V assigns T or F to $U_i$. Then B is a theorem of PS*. But B is false for every interpretation. Therefore $\sim$B is true for every interpretation; i.e. $\sim$B is logically valid. So by the semantic completeness of PS $\sim$B is a theorem of PS and therefore also of PS*. So both B and $\sim$B are theorems of PS*; i.e. PS* is inconsistent. But U was *any* schema unprovable in PS. So PS is syntactically complete.

Though no unprovable *schema* can consistently be added as an axiom-schema to PS, some unprovable *formulas* can consistently be added as axioms:

33.2 *If $\sim A$ is any formula of P that is not a theorem of PS, then A can consistently be added to PS as an axiom*

*Proof.* Let $\sim$A be any formula of P that is not a theorem of PS. Suppose the addition of A as an axiom to PS yields an inconsistent system. Then for some formula B both $A \vdash_{PS} B$ and $A \vdash_{PS} \sim B$. So, by the Deduction Theorem, $\vdash_{PS} A \supset B$ and $\vdash_{PS} A \supset \sim B$, for some B. But by 32.4 $\vdash_{PS} (A \supset B) \supset ((A \supset \sim B) \supset \sim A)$. So by MP twice $\vdash_{PS} \sim A$. But this contradicts our assumption that $\sim$A was not a theorem of PS. So: If $\sim$A is any formula of P that is not a theorem of PS, A can consistently be added to PS as an axiom.

*Remarks*

1. A formal system of truth-functional propositional logic for which simple and absolute consistency do not coincide may be semantically and syntactically complete, and yet it may be possible to add an unprovable schema without getting *absolute* inconsistency. This is true of the following system, due to Henry Hiż (1957):

*Hiż's system*

Symbols and formulas: As for PS, with $\sim$ and $\supset$ the sole connectives

Axiom-schemata:

1. $\sim(A \supset B) \supset A$
2. $\sim(A \supset B) \supset \sim B$

Rules of inference:

[Notice the requirement that the premisses must be theorems]

1. If $A \supset B$ and $B \supset C$ are theorems, so is $A \supset C$.
2. If $A \supset (B \supset C)$ and $A \supset B$ are theorems, so is $A \supset C$.
3. If $\sim A \supset B$ and $\sim A \supset \sim B$ are theorems, so is A.

All tautologies of P are theorems of this system, so it is syntactically complete (cf. the proof of 33.1), but it is possible to add to the system the unprovable schema $\sim A$ without getting *absolute* inconsistency: e.g. no unnegated propositional symbol is a theorem of the enlarged system.

2. If a system has a (standard) substitution rule as one of its rules of inference, then adding to it as an axiom a formula that is not a theorem amounts to the same thing as adding to it a (previously) unprovable *schema*, for the rule of substitution will permit the substitution of arbitrary formulas for the propositional symbols of the added formula. So if such a system is syntactically complete, *no formula* will be consistently addable (i.e. no version of 33.2 will hold for it). In this respect there is a fundamental difference between syntactically complete systems *with* and those (like PS) *without* substitution as a rule of inference.

## 34 Proof of the decidability of PS. Decidable system and decidable formula. Definition of *effective proof procedure*

*Definition.* A system S is *decidable* iff there is an effective method for telling, for each formula of S, whether or not it is a theorem of S.

34.1 *PS is decidable*

*Proof.* By 28.3 every theorem of PS is a tautology of P, and by the semantic completeness theorem for PS every tautology of P is a theorem of PS. So a formula of P is a theorem of PS iff it is a tautology of P. We take it as obvious that the complete truth-table method is an effective method for telling, for any formula of P, whether or not it is a tautology of P. So PS is decidable.

*Note.* Though PS and other formal systems of the full standard truth-functional propositional logic are decidable, it has been proved of some 'fragments' of standard truth-functional logic that they are undecidable. For details and references, and for other twists in the metatheory of standard truth-functional propositional logic, see the interesting survey paper, Harrop (1964).

Though the *system* PS is decidable, there are *formulas* of PS that are not decidable in PS, in a different sense of 'decidable' to be explained immediately:

*Definition.* A *formula* A is *decidable in a system* S iff either A or its negation is a theorem of S.

Examples: The formula $\sim(p' \supset p')$ is decidable in PS; the formula $\sim p'$ is not.

So a *decidable system* may have *undecidable formulas* (PS is such a system). Conversely, *every formula* in an *undecidable system* may be *decidable*. Hint for an example: Let S be a consistent undecidable system, with negation, whose formulas are effectively enumerable. By using Lindenbaum's Lemma and its proof, we can define a system S′ that is got by adding to S successively as axioms each formula of S that can consistently be added when its number comes up (so to speak), so that the set of theorems of S′ is a maximal p-consistent set of S. Then for each formula A of S either A or its negation will be a theorem of S′. But nothing in the way that S′ was defined or constructed need guarantee that there is an *effective method* for telling, for each formula A, *which* of the two formulas, A and the negation of A, is a theorem of S′.

*Definition.* A system S has an *effective proof procedure* iff, given any arbitrary theorem T of S, there is an effective method for constructing a proof in S of T.

A decidable system need not have an effective proof procedure. In fact PS does have an effective proof procedure. The

Kalmár-type completeness proof can be made to yield one, but it is rather clumsy. We shall give an example of a much simpler effective proof procedure for a different system of truth-functional propositional logic in §37.

## 35 Extended sense of 'interpretation of P'. Finite weak models and finite strong models

In this section we widen the notion of 'interpretation of P':

1. We shall allow as interpretations of P assignments of values other than truth values to the propositional symbols of P.

2. We shall allow $\sim$ and $\supset$ to be defined by tables that are not truth tables, though they will in fact bear a close resemblance to truth tables.

3. Where before 'all interpretations' meant 'all interpretations assigning values exclusively from the set {T, F} and giving $\sim$ and $\supset$ their usual senses', we shall now want to speak of 'all interpretations assigning values from such and such a set and with such and such senses of $\sim$ and $\supset$', where the set and the senses will vary.

4. 'Logically valid' was defined as meaning 'true for all interpretations' where that meant 'true for all interpretations assigning values exclusively from the set {T, F} and giving $\sim$ and $\supset$ their usual senses'. It will still mean this (i.e. the same as the long phrase). But we shall also make use of the word 'valid' in another way. We shall use it to mean 'true for all interpretations belonging to the class of interpretations under consideration', where the main things determining what the class is will be (1) the particular set of values that interpretations in the class may draw on for their assignments to the propositional symbols, and (2) the particular senses given to $\sim$ and $\supset$ (constant for a given class of interpretations).

So: A *class* M of interpretations of P will be given by specifying or exhibiting:

1. A non-empty set V of things called *values*, the values that may be assigned to the propositional symbols of P.

2. A subset D of V, the set of *designated values.*

3. (i) A table T1 showing, for an arbitrary formula A of P,

what value (from V) $\sim$A has for each value that may be assigned to A from the set V.

(ii) A table T2 showing what value A $\supset$ B has for each possible pair of values from V that A and B may have (A and B being arbitrary formulas of P).

To illustrate: For the class of interpretations in earlier sections

1. V was the set {T, F}, i.e. the set of *truth* values.
2. D was the set {T}, i.e. the set whose sole member is the truth value truth.
3. (i) The table T1 was the standard truth table for $\sim$.

(ii) The table T2 was the standard truth table for $\supset$.

Let there be given a class M of interpretations of P, for which the sets and tables V, D, T1 and T2 have been specified. We define next an *interpretation* I as an assignment to each of the propositional symbols of P of one or other of the values in the set V. A formula A is *true for I* iff it takes a designated value (i.e. a value from the set D) for the assignment of values that I makes to its constituent propositional symbols (the value of A being determined by this assignment and the tables T1 and T2). A formula A is *valid for the class M* iff it is *true for all interpretations of the class M*.

We shall make use of these notions in the independence proofs in the next section.

The following notions [due to Ronald Harrop: cf. Harrop, 1964] belong here too. We shall not be making any use of them. But they are to be encountered in the literature, and they would be useful in a more extended treatment than ours. We define the notions only for the language P; but they could be defined for arbitrary languages.

An interpretation I belonging to a class M (for which V, D, T1 and T2 have been specified) is said to be a *finite weak model* of an arbitrary formal system S with language P if it satisfies conditions 1, 2 and 3*a* below, and a *finite strong model* if it satisfies conditions 1, 2 and 3*b* below:

1. The set V is finite.
2. Every axiom of S is valid for the class M, i.e. true for all interpretations of the class M.

3*a*. Each rule of inference of S preserves validity-for-the-class-M.

3*b*. Each rule of inference of S preserves truth for a given interpretation.

To illustrate: Every interpretation of P belonging to the class of interpretations under consideration in earlier sections is both a finite weak model and a finite strong model of the system PS. For (1) the set {T, F} is finite; (2) the axioms of PS are true for all interpretations of that class; (3*a*) Modus Ponens for ⊃ preserves truth-for-all-interpretations-in-that-class [28.2]; and (3*b*) Modus Ponens preserves truth-for-I [28.5].

## 36 Proof of the independence of the three axiom-schemata of PS

Let S be any formal system, and let A be one of its axioms. Let S – A be the system got from S by omitting A from the axioms. Then A is *independent* of the other axioms of S iff there is no proof in S – A of A [i.e. iff A is not a theorem of S – A]. Similarly, with obvious modifications, for axiom-schemata.

It is easy to see how in principle one can establish that an axiom is *not* independent of the other axioms of the system (viz. by proving it from the others). But it is perhaps not so obvious how in principle to establish independence. In fact, both the sorts of method used in proving consistency may be used in proving independence:

(*a*) Model-theoretic. Let A be the axiom whose independence we want to prove. If we can show that there is an interpretation for which all the other axioms come out true, and for which the rule(s) of inference preserve(s) truth, but for which A does not come out true, then this proves that A is not derivable from the other axioms, i.e. that A is independent of the other axioms.

(*b*) Proof-theoretic. Again let A be the axiom we are interested in. If we can show that all the other axioms have a certain syntactic property, that this property is transmitted by the rule(s) of inference, and that A does not have that property, then A is independent of the other axioms.

If we restricted ourselves to the earlier, narrow sense of 'interpretation' we could not use Method (*a*) to prove the independence of the axioms of PS. For in that old narrow sense of 'interpretation' every axiom of PS is true for every interpretation. But we shall see that in the new extended sense of 'interpretation' it is possible to find interpretations for which some of the axioms do not come out true.

In what follows we shall speak freely of schemata being 'valid' or 'not valid'. This is simply an abbreviated way of saying that all formulas of the form of the schema are valid, or that not all formulas of that form are valid, respectively. Similarly for 'provable'.

36.1 *The axiom-schema PS 1 is independent of the set of axiom-schemata {PS 2, PS 3}*

*Proof.* Let M be the class of interpretations of P with V = {0, 1, 2}, D = {0}, and T1 and T2 as follows:

T1

| A | ~A |
|---|---|
| 0 | 1 |
| 1 | 1 |
| 2 | 0 |

T2

| A | B | A ⊃ B |
|---|---|---|
| 0 | 0 | 0 |
| 0 | 1 | 2 |
| 0 | 2 | 2 |
| 1 | 0 | 2 |
| 1 | 1 | 2 |
| 1 | 2 | 0 |
| 2 | 0 | 0 |
| 2 | 1 | 0 |
| 2 | 2 | 0 |

The tables below show that the axiom-schemata PS 2 and PS 3 are valid for all interpretations in the class M. Modus Ponens preserves validity for the class M. Therefore anything provable using only axioms by PS 2 and PS 3 is valid for the class M. But the schema PS 1 is not valid for the class M. Therefore it is not provable from the set {PS 2, PS 3}. (Each table needs $3^n$ rows where $n$ is the number of distinct schematic letters in the axiom-schema.)

| PS 1 | PS 2 | PS 3 |
|---|---|---|
| A ⊃ (B ⊃ A) | (A ⊃ (B ⊃ C)) ⊃ ((A ⊃ B) ⊃ (A ⊃ C)) | (~A ⊃ ~B) ⊃ (B ⊃ A) |
| 0 0 0 0 0 | 0 0 0 0 0 0 0 0 0 0 0 0 0 | 1 0 2 1 0 0 0 0 0 |
| 0 2 1 2 0 | 0 2 0 2 1 0 0 0 0 2 0 2 1 | 1 0 2 1 1 0 1 2 0 |
| 0 0 2 0 0 | 0 2 0 2 2 0 0 0 0 2 0 2 2 | 1 0 2 0 2 0 2 0 0 |
| 1 0 0 2 1 | 0 2 1 2 0 0 0 2 1 0 0 0 0 | 1 1 2 1 0 0 0 2 1 |
| 1 0 1 2 1 | 0 2 1 2 1 0 0 2 1 0 0 2 1 | 1 1 2 1 1 0 1 2 1 |

[Tables continued on p. 124]

| PS1 | PS2 | PS3 |
|---|---|---|
| A ⊃ (B ⊃ A) | (A ⊃ (B ⊃ C)) ⊃ ((A ⊃ B) ⊃ (A ⊃ C)) | ( ~ A ⊃ ~ B) ⊃ (B ⊃ A) |
| 1 2 2 0 1 | 0 0 1 0 2 0 0 2 1 0 0 2 2 | 1 1 2 0 2 0 2 0 1 |
| 2 0 0 2 2 | 0 0 2 0 0 0 0 2 2 0 0 0 0 | 0 2 2 1 0 0 0 2 2 |
| 2 0 1 0 2 | 0 0 2 0 1 0 0 2 2 0 0 2 1 | 0 2 2 1 1 0 1 0 2 |
| 2 0 2 0 2 | 0 0 2 0 2 0 0 2 2 0 0 2 2 | 0 2 0 0 2 0 2 0 2 |
| * | 1 2 0 0 0 0 1 2 0 0 1 2 0 | * |
| | 1 0 0 2 1 0 1 2 0 0 1 2 1 | |
| | 1 0 0 2 2 0 1 2 0 0 1 0 2 | |
| | 1 0 1 2 0 0 1 2 1 0 1 2 0 | |
| | 1 0 1 2 1 0 1 2 1 0 1 2 1 | |
| | 1 2 1 0 2 0 1 2 1 0 1 0 2 | |
| | 1 2 2 0 0 0 1 0 2 2 1 2 0 | |
| | 1 2 2 0 1 0 1 0 2 2 1 2 1 | |
| | 1 2 2 0 2 0 1 0 2 0 1 0 2 | |
| | 2 0 0 0 0 0 2 0 0 0 2 0 0 | |
| | 2 0 0 2 1 0 2 0 0 0 2 0 1 | |
| | 2 0 0 2 2 0 2 0 0 0 2 0 2 | |
| | 2 0 1 2 0 0 2 0 1 0 2 0 0 | |
| | 2 0 1 2 1 0 2 0 1 0 2 0 1 | |
| | 2 0 1 0 2 0 2 0 1 0 2 0 2 | |
| | 2 0 2 0 0 0 2 0 2 0 2 0 0 | |
| | 2 0 2 0 1 0 2 0 2 0 2 0 1 | |
| | 2 0 2 0 2 0 2 0 2 0 2 0 2 | |
| | * | |

36.2 *The schema PS 2 is independent of the set {PS 1, PS 3}*

*Proof.* Let M be the class of interpretations of P with V ={0, 1, 2}, D ={1}, and T1 and T2 as follows:

T1

| A | ~A |
|---|---|
| 0 | 1 |
| 1 | 0 |
| 2 | 2 |

T2

| A | B | A ⊃ B |
|---|---|---|
| 0 | 0 | 1 |
| 0 | 1 | 1 |
| 0 | 2 | 1 |
| 1 | 0 | 0 |
| 1 | 1 | 1 |
| 1 | 2 | 2 |
| 2 | 0 | 2 |
| 2 | 1 | 1 |
| 2 | 2 | 1 |

PS 1 and PS 3 are valid for the class M. MP preserves validity for the class M. PS 2 is not valid for the class M (it gets the value 2 when A gets 2, B gets 2, and C gets 0).

36.3 *PS 3 is independent of the set {PS 1, PS 2}*

*Proof.* We use a different method of proof, for variety. Let L* be the schema that results from a schema L by deleting all negation signs in L. Thus, if L is $(\sim A \supset \sim B)$, then L* is $(A \supset B)$. Call L* the *associated schema* of L. Then (1) the associated schemata of PS 1 and PS 2 are both tautological schemata [in both cases L* = L]; (2) Modus Ponens preserves the tautologicality of associated schemata [notice that $(A \supset B)^*$ is $(A^* \supset B^*)$]; (3) the associated schema of PS 3 is *not* a tautological schema [it is $(A \supset B) \supset (B \supset A)$]. So PS 3 is independent of {PS 1, PS 2}.

## 37 Anderson and Belnap's formalisation of truth-functional propositional logic: the system AB

Modus Ponens for $\supset$ can be written like this:

$$\frac{A, \quad A \supset B}{B}$$

$A \supset B \equiv \sim A \vee B$. So Modus Ponens for $\vee$ is:

$$\frac{A, \quad \sim A \vee B}{B}$$

Alan Ross Anderson and Nuel Belnap object to Modus Ponens for $\vee$ (and accordingly also to Modus Ponens for $\supset$) for the following reason: they hold that

(*a*) the rules of inference of a formal system whose theorems (on their intended interpretation) are truths of logic ought, on their intended interpretation, to be valid rules of inference;

(*b*) on its intended interpretation Modus Ponens for $\vee$ (or for $\supset$) implies (in conjunction with accepted principles) that from any arbitrary formal contradiction there follows logically any proposition you like to take; and

(*c*) it is just not true that from any arbitrary formal contradiction there follows logically any proposition you like to take.

To illustrate (*b*):

| | | |
|---|---|---|
| 1 | A | [Assumption] } The formal contradiction |
| 2 | $\sim A$ | [Assumption] } |
| 3 | $\sim A \vee B$ | [From 2 by accepted principles of truth-functional logic: B may be any arbitrary formula] |
| 4 | B | [1, 3, MP for $\vee$] |

Accordingly Anderson and Belnap presented (1959) a formal system of truth-functional propositional logic that did not employ Modus Ponens for $\vee$ (or any equivalent of it). This system, which is a simplification of earlier systems of Kurt Schütte, admits of a very easy completeness proof and has a simple effective proof procedure.

[*Philosophical comments.* I agree with Anderson and Belnap about (*c*). But I do not think that *for our present purpose* we have to agree with (*a*). I regard the rule of inference of PS, not as something that on its intended interpretation has to be a valid rule of inference, but merely as a *rule for generating formulas* that, on their intended interpretation, express truths of logic. It does not matter for our present purpose *how* these formulas are generated, provided that they are all generated (and that nothing else is). So for us either Modus Ponens has no intended interpretation, or it is to be interpreted as a rule for generating formulas from other formulas, a rule that, given the interpretation of the formulas, turns out to generate only truths of logic when applied only to truths of logic. Accordingly I deny (*b*). However I agree with Anderson and Belnap's implied claim that *one* of the logician's tasks is to construct formal systems whose rules of inference can be interpreted as expressing valid rules of inference.]

### *The system AB*

*Symbols*

AB has just 4 symbols, viz. $p \quad ' \quad \overline{\phantom{p}} \quad \vee$

The symbol $p$ followed by one or more dashes is a *propositional symbol* of AB.

The bar, $\overline{\phantom{p}}$, on the intended interpretation expresses negation. Its use enables us to dispense with brackets.

*Wffs*

1. Any propositional symbol is a wff.
2. If A is a wff, then $\bar{\mathrm{A}}$ is a wff.
3. If A and B are wffs, then A $\vee$ B is a wff.
4. Nothing else is a wff.

*Definition* of *primitive disjunction*:

A wff A is a *primitive disjunction* iff it has the form $\mathrm{B}_1 \vee \ldots \vee \mathrm{B}_n$ $(n \geqslant 1)$ where each $\mathrm{B}_i$ is either a propositional symbol or a propositional symbol with the bar (—) over it.

*Definition* of *disjunctive part*:

(*a*) Every wff is a disjunctive part of itself.
(*b*) If B $\vee$ C is a disjunctive part of A, then B is a disjunctive part of A and so is C.

*Notation.* D(A) is a wff of which A is a disjunctive part, and D(B) is the result of replacing A in D(A) by B.

*Axioms*

A wff A is an axiom iff it is a primitive disjunction and, for some propositional symbol B, both B and $\bar{\mathrm{B}}$ are disjunctive parts of it.

Examples: $p' \vee p'' \vee \overline{p'''} \vee \overline{p'}$ is an axiom; $p' \vee \overline{p'} \vee \overline{p'' \vee p'''}$ is not (because it is not a primitive disjunction).

*Rules of inference*

I. $\mathrm{D}(\bar{\bar{\mathrm{A}}})$ is an immediate consequence of D(A).
II. $\mathrm{D}(\overline{\mathrm{A} \vee \mathrm{B}})$ is an immediate consequence of $\mathrm{D}(\bar{\mathrm{A}})$ and $\mathrm{D}(\bar{\mathrm{B}})$.

For legibility from now on we shall write *p*, *q*, *r* instead of $p'$, $p''$, $p'''$.

Examples:

I. $p \vee \overline{\overline{\overline{q \vee r}}}$ is an immediate consequence of $p \vee \overline{q \vee r}$.
II. $p \vee \overline{q \vee r}$ is an immediate consequence of $p \vee \bar{q}$ and $p \vee \bar{r}$.

*Definition* of *proof in AB*:

A *proof in AB* is a finite string of formulas of AB each one of which is either an axiom of AB or an immediate consequence by Rule I or Rule II of some formula(s) preceding it in the string.

Example:

| | | |
|---|---|---|
| 1 | $\bar{p}\vee\bar{r}\vee\bar{p}\vee q$ | [Axiom] |
| 2 | $\bar{\bar{p}}\vee\bar{r}\vee\bar{p}\vee q$ | [From 1 by Rule I] |
| 3 | $\bar{q}\vee\bar{r}\vee\bar{p}\vee q$ | [Axiom] |
| 4 | $\overline{\bar{p}\vee q}\vee\bar{r}\vee\bar{p}\vee q$ | [From 2 and 3 by Rule II] |

One beauty of this system is that there is a simple effective proof procedure for it. To illustrate:

Required to find a proof of $\overline{\bar{\bar{p}}\vee\bar{q}}\vee\bar{q}\vee p$ [which is a transcription into — and $\vee$ of $(\sim p\supset\sim q)\supset(q\supset p)$].

1. Look at the leftmost unreduced part of the formula to be proved [a 'reduced' part is a disjunctive part that is either a propositional symbol or a negated propositional symbol]. In this case it is

$$\overline{\bar{\bar{p}}\vee\bar{q}}$$

The only rule that can give us anything of this form is Rule II, which says that

$$\overline{\bar{\bar{p}}\vee\bar{q}}\vee\bar{q}\vee p$$

is obtainable from the two formulas

$$\bar{\bar{\bar{p}}}\vee\bar{q}\vee p$$

and

$$\bar{\bar{q}}\vee\bar{q}\vee p$$

2. Begin the construction of a tree, with the formula to be proved at the bottom, and the two formulas just mentioned at the ends of the first fork:

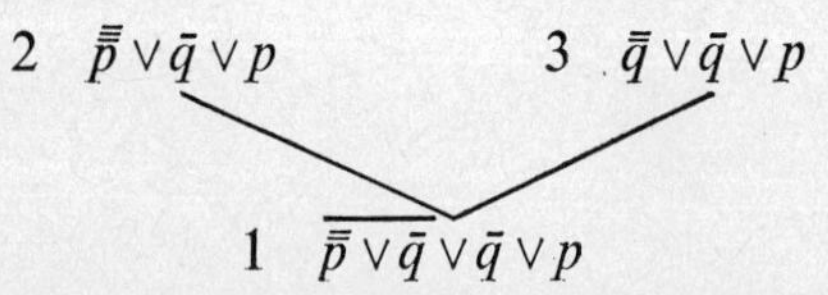


3. Look at the leftmost unreduced part of $\bar{\bar{\bar{p}}}\vee\bar{q}\vee p$ [2]. The only rule that can give us anything of this form is Rule I, which says that

$$\bar{\bar{\bar{p}}}\vee\bar{q}\vee p$$

is obtainable from

$$\bar{p}\vee\bar{q}\vee p$$

4. Write in $\bar{p}\vee\bar{q}\vee p$ above $\bar{\bar{\bar{p}}}\vee\bar{q}\vee p$ in the tree. Our tree now looks like this:

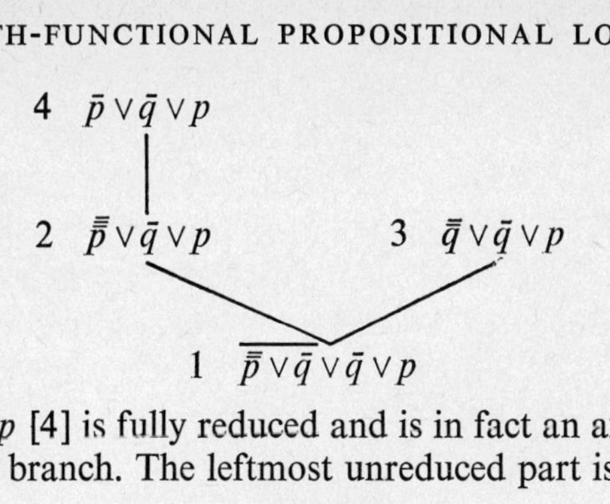


5. $\bar{p} \vee \bar{q} \vee p$ [4] is fully reduced and is in fact an axiom. Look to the other branch. The leftmost unreduced part is

$$\bar{\bar{q}}$$

obtainable only by Rule I, which says that

$$\bar{\bar{q}} \vee \bar{q} \vee p$$

is obtainable from

$$q \vee \bar{q} \vee p$$

6. Write in $q \vee \bar{q} \vee p$ above $\bar{\bar{q}} \vee \bar{q} \vee p$ [3]. This gives us the tree:

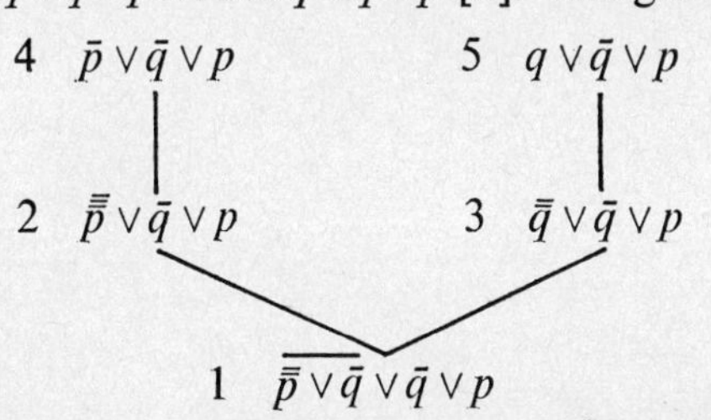


7. Both branches of the tree end in fully reduced formulas that are in fact axioms. In order to get the required proof we simply write down the formulas in the tree in reverse order, thus:

| | | |
|---|---|---|
| [5] | $q \vee \bar{q} \vee p$ | [Axiom] |
| [4] | $\bar{p} \vee \bar{q} \vee p$ | [Axiom] |
| [3] | $\bar{\bar{q}} \vee \bar{q} \vee p$ | [From 5 by Rule I] |
| [2] | $\bar{\bar{p}} \vee \bar{q} \vee p$ | [From 4 by Rule I] |
| [1] | $\overline{\bar{p} \vee \bar{q}} \vee \bar{q} \vee p$ | [From 2 and 3 by Rule II] |

It is intuitively evident that the method we have been using is an effective method for constructing, for any given formula, a tree with a finite number of branches each of which ends in a primitive disjunction (counting a single formula as a tree with zero branches). For it is impossible for a formula of AB to be neither a primitive disjunction nor further reducible by our method (if it is not a primitive disjunction, then we can reduce it at least one stage further). We leave it to the reader to give a rigorous proof of this by mathematical induction.

It is obvious also that if every branch of such a tree ends in an axiom, the corresponding formula is a theorem of AB. The converse of this statement is a consequence of the proof of the semantic completeness of AB, which follows the proof of consistency.

*Semantics of AB*

As for P, with obvious modifications. E.g. clause 3 of the definition of *true for an interpretation of P* is replaced by

3. $A \vee B$ is true for I iff either A is true for I or B is true for I [truth-functional 'either . . . or . . .'].

*Proof of the consistency of AB*

37.1 *Lemma: The rules of inference of AB preserve tautologyhood*

*Proof*

(*a*) Rule I. To prove that if D(A) is a tautology, then so is $D(\bar{\bar{A}})$.

Cases: 1. D(A) is A.
2. D(A) is $A \vee B$.
3. D(A) is $B \vee A$.
4. D(A) is $B \vee A \vee C$.

Case 1. It follows from the definition of truth for AB that if $\models_{AB} A$, then $\models_{AB} \bar{\bar{A}}$.

Case 2. If $\models_{AB} A \vee B$, then $\models_{AB} \bar{\bar{A}} \vee B$.

Cases 3 and 4. Similar to Case 2.

(*b*) Rule II. To prove that if $\models_{AB} D(\bar{A})$ and $\models_{AB} D(\bar{B})$, then $\models_{AB} (\overline{A \vee B})$.

Cases: 1. $D(\bar{A})$ is $\bar{A}$ [so $D(\bar{B})$ is $\bar{B}$ and $D(A \vee B)$ is $\overline{A \vee B}$].
2. $D(\bar{A})$ is $\bar{A} \vee C$ [so $D(\bar{B})$ is $\bar{B} \vee C$ and $D(\overline{A \vee B})$ is $\overline{A \vee B} \vee C$].
3. $D(\bar{A})$ is $C \vee \bar{A}$.
4. $D(\bar{A})$ is $C \vee \bar{A} \vee D$.

Case 1. It follows from the definition of truth for AB that if $\models_{AB} \overline{A}$ and $\models_{AB} \overline{B}$, then $\models_{AB} \overline{A \vee B}$.

Case 2. From the equivalence

$$((\overline{A} \vee C) \wedge (\overline{B} \vee C)) \equiv (\overline{A \vee B} \vee C)$$

it is clear that if $\models_{AB} \overline{A} \vee C$ and $\models_{AB} \overline{B} \vee C$, then $\models_{AB} \overline{A \vee B} \vee C$.

Cases 3 and 4. Similar to Case 2.

37.2 *AB is consistent*

*Proof.* Every axiom of AB is a tautology, and by 37.1 the rules of inference preserve tautologyhood. So every theorem of AB is a tautology. So AB is consistent.

*Proof of the semantic completeness of AB*

37.3 *Lemma: Any formula that is an immediate consequence of a formula that is false for a given interpretation is also false for that interpretation*

[This means, in the case of Rule II, that if *either* D($\overline{A}$) or D($\overline{B}$) is false for the interpretation (or both are), then D($\overline{A \vee B}$) is also false for it]

*Proof*

(*a*) Rule I. To prove that if D(A) is false for I then so is D($\overline{\overline{A}}$).

Cases 1–4 are as in the proof for (*a*) of 37.1. The proofs for all these cases are obvious.

(*b*) Rule II.

Case 1. D($\overline{A}$) is $\overline{A}$. So D($\overline{B}$) is $\overline{B}$, and D($\overline{A \vee B}$) is $\overline{A \vee B}$. If $\overline{A}$ is false for I, then $\overline{A \vee B}$ is false for I, no matter whether $\overline{B}$ is true or false for I. Similarly for $\overline{B}$.

Case 2. D($\overline{A}$) is $\overline{A} \vee C$. So D($\overline{B}$) is $\overline{B} \vee C$, and D($\overline{A \vee B}$) is $\overline{A \vee B} \vee C$. If $\overline{A} \vee C$ is false for I, then A must be true for I and C must be false for I, and so $\overline{A \vee B} \vee C$ must be false for I. Similarly for $\overline{B} \vee C$.

Cases 3 and 4. Similar to 2.

37.4 *AB is semantically complete* [*with respect to tautologies in* $\overline{\phantom{A}}$ *and* $\vee$ ]

*Proof.* Suppose A is a formula of AB that is not a theorem of

AB. Construct a tree for A in the way illustrated near the beginning of this section. At least one branch of the tree must end in a primitive disjunction that is not an axiom [if every branch ended in an axiom, then A would be a theorem]. Call any such branch a 'bad' branch. For a given bad branch, the following interpretation I of AB makes every formula in the branch false for I:

> Let $B_1, \ldots, B_k$ be all the distinct propositional symbols occurring in the formula at the top of the bad branch. For each $B_i$ $(1 \leqslant i \leqslant k)$, if $B_i$ occurs unnegated there, I assigns F to it; if negated, T [for no $B_i$ do both $B_i$ and $\overline{B}_i$ occur there; otherwise the formula would be an axiom]. I assigns T to every other propositional symbol in AB [it could just as well assign them F, or any other combination of truth values].

On this interpretation the formula at the top of the bad branch comes out false, and by 37.3 falsity for I is transmitted by the rules of inference. So *every* formula in the bad branch, including A at the bottom, is false for I. So A is not a tautology. So we have: If A is a formula of AB that is not a theorem of AB, then A is not a tautology. Or equivalently: If A is a tautology of AB, then A is a theorem of AB.

37.5 *AB is decidable*

2 proofs:

1. Since a formula of AB is a theorem of AB iff it is a tautology, the usual truth table method is an effective decision method for AB.

2. Let A be any formula of AB for which we want to know whether or not it is a theorem. Construct a (complete) tree for A in the way illustrated earlier. If every branch of the tree ends in an axiom, A is a theorem of AB. If at least one branch ends in a formula that is not an axiom, then A is not a theorem.

37.6 *There is an effective proof procedure for AB*

*Proof.* Let A be any theorem of AB. Construct a tree for A in the usual way. Take the formulas in the tree in the reverse order from the order in which they occurred in the construction of the tree. The result is a proof in AB of A.

To end this section we give a proof tree for, and a proof of, the formula

$$\overline{\bar{p} \vee \bar{q} \vee r} \vee \overline{\bar{p} \vee q} \vee \bar{p} \vee r$$

which is a transcription into $\overline{\phantom{x}}$ and $\vee$ of the formula

$$(p \supset (q \supset r)) \supset ((p \supset q) \supset (p \supset r))$$

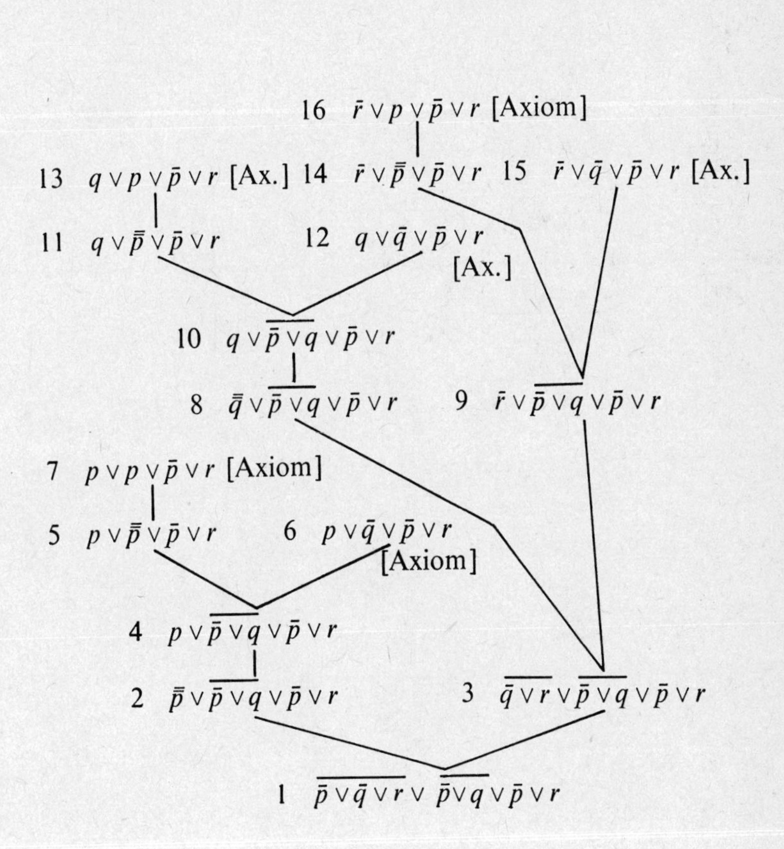

Proof in AB of $\overline{\bar{p} \vee \bar{q} \vee r} \vee \overline{\bar{p} \vee q} \vee \bar{p} \vee r$ [cf. the proof tree]:

| | | |
|---|---|---|
| 16 | $\bar{r} \vee p \vee \bar{p} \vee r$ | [Axiom] |
| 15 | $\bar{r} \vee \bar{q} \vee \bar{p} \vee r$ | [Axiom] |
| 14 | $\bar{r} \vee \bar{\bar{p}} \vee \bar{p} \vee r$ | [16, Rule I] |
| 13 | $q \vee p \vee \bar{p} \vee r$ | [Axiom] |
| 12 | $q \vee \bar{q} \vee \bar{p} \vee r$ | [Axiom] |
| 11 | $q \vee \bar{\bar{p}} \vee \bar{p} \vee r$ | [13, Rule I] |
| 10 | $q \vee \overline{\bar{p} \vee q} \vee \bar{p} \vee r$ | [11, 12, Rule II] |
| 9 | $\bar{r} \vee \overline{\bar{p} \vee q} \vee \bar{p} \vee r$ | [14, 15, II] |
| 8 | $\bar{\bar{q}} \vee \overline{\bar{p} \vee q} \vee \bar{p} \vee r$ | [10, I] |
| 7 | $p \vee p \vee \bar{p} \vee r$ | [Axiom] |
| 6 | $p \vee \bar{q} \vee \bar{p} \vee r$ | [Axiom] |
| 5 | $p \vee \bar{\bar{p}} \vee \bar{p} \vee r$ | [7, I] |
| 4 | $p \vee \overline{\bar{p} \vee q} \vee \bar{p} \vee r$ | [5, 6, II] |
| 3 | $\overline{\bar{q} \vee r} \vee \overline{\bar{p} \vee q} \vee \bar{p} \vee r$ | [8, 9, II] |
| 2 | $\bar{\bar{p}} \vee \overline{\bar{p} \vee q} \vee \bar{p} \vee r$ | [4, I] |
| 1 | $\overline{\bar{p} \vee \bar{q} \vee r} \vee \overline{\bar{p} \vee q} \vee \bar{p} \vee r$ | [2, 3, II] |

PART THREE

# First Order Predicate Logic: Consistency and Completeness

## 38 A formal language for first order predicate logic: the language Q. The languages Q+

*The language Q*
('Q' for 'Quantification')

*Symbols*

$p \quad ' \quad x \quad a \quad f \quad F \quad * \quad \sim \quad \supset \quad \Lambda \quad ( \quad )$

Names for various combinations of these symbols:

*Propositional symbols*: $p', p'', p''', \ldots$
*Individual variables* (*variables*, for short): $x', x'', x''', \ldots$
*Individual constants* (*constants*, for short): $a', a'', a''', \ldots$
*Function symbols*: $f^{*\prime}, f^{*\prime\prime}, f^{*\prime\prime\prime}, \ldots, f^{**\prime}, f^{**\prime\prime}, f^{**\prime\prime\prime}, \ldots,$
$f^{***\prime}, f^{***\prime\prime}, f^{***\prime\prime\prime}, \ldots, \ldots$

i.e. the lower-case italic letter *f* followed by one or more asterisks and then one or more dashes is a function symbol.

A function symbol with exactly *n* asterisks is an *n-place function symbol.*

*Predicate symbols*: $F^{*\prime}, F^{*\prime\prime}, F^{*\prime\prime\prime}, \ldots, F^{**\prime}, F^{**\prime\prime}, F^{**\prime\prime\prime}, \ldots,$
$F^{***\prime}, F^{***\prime\prime}, F^{***\prime\prime\prime}, \ldots, \ldots$

i.e. the capital italic letter *F* followed by one or more asterisks and then one or more dashes is a predicate symbol. A predicate symbol with exactly *n* asterisks is an *n-place predicate symbol.*

*Connectives*: $\sim \quad \supset$
*Universal quantifier*: $\Lambda$
*Brackets*: ( )
*Terms*: An individual constant is a term. An individual variable is a term. An *n*-place function symbol followed by *n* terms is a term. Nothing else is a term.
*Closed terms*: A term is closed iff no variable occurs in it.

*Wffs*

1. Any propositional symbol is a wff and an atomic wff.
2. If F is an *n*-place predicate symbol and $t_1, \ldots, t_n$ are terms

(not necessarily distinct), then $Ft_1 \ldots t_n$ is a wff and an atomic wff.

3. If A is a wff and v an individual variable, then $\Lambda$vA is a wff.
4. If A is a wff, then $\sim$A is a wff.
5. If A and B are wffs, then $(A \supset B)$ is a wff.
6. Nothing else is a wff.

*Scope*

In $\Lambda$vA if A is a wff then A is the scope of the quantifier $\Lambda$.

Examples:

1. In $\Lambda x'(F^{*\prime}x' \supset F^{*\prime\prime}x'')$ the scope of the quantifier is $(F^{*\prime}x' \supset F^{*\prime\prime}x'')$.

2. In $(\Lambda x'F^{*\prime}x' \supset F^{*\prime\prime}x'')$ the scope of the quantifier is $F^{*\prime}x'$: for $F^{*\prime}x' \supset F^{*\prime\prime}x'')$ is not a wff.

*Free and bound occurrences of variables*

An occurrence of a variable v is *bound* in a wff iff either it immediately follows a quantifier[1] in the wff, or it is within the scope of a quantifier[1] that has v as its variable (i.e. a quantifier[1] that is immediately followed by v). Otherwise an occurrence is *free* in the wff.

A variable is free in a wff if any occurrence of it is free in the wff, and a variable is bound in a wff if any occurrence of it is bound in the wff. (So a variable may be both bound and free in the same wff.)

Examples:

1. In $\Lambda x'(F^{*\prime}x' \supset F^{*\prime\prime}x'')$ both occurrences of $x'$ are bound, while the occurrence of $x''$ is free.

2. In $(\Lambda x'F^{*\prime}x' \supset F^{*\prime}x')$ the first two occurrences of $x'$ are bound while the third occurrence is free. So the variable $x'$ is both bound and free in this formula.

It follows from our definitions that any occurrence of a variable in an *atomic* wff (i.e. a wff without quantifiers) is free.

*Notation*

If A is a wff, t a term, v a variable, then At/v is the wff got from A by substituting t for all free occurrences of v in A.

[1] i.e. an occurrence of the quantifier. Similarly elsewhere.

Examples:

1. If A is $(\bigwedge x' F^{*\prime} x' \supset F^{*\prime} x')$, then $\mathrm{A}x'''/x'$ is $(\bigwedge x' F^{*\prime} x' \supset F^{*\prime} x''')$.

2. If A is $\bigwedge x'(F^{*\prime} x' \supset F^{*\prime} x')$, then $\mathrm{A}x'''/x'$ is A itself, since there is no free occurrence of $x'$ in A.

*Definition. t is free for v in A*

*t is free for v in A* if (1) if t is a variable, then t occurs free in At/v wherever v occurs free in A, and (2) if t is a term in which any variables occur, then wherever t is substituted for free occurrences of v in A all occurrences of variables in t remain free. If t is closed term, then t is free from v in any formula A.

Examples:

1. In $(\bigwedge x' F^{*\prime} x' \supset F^{*\prime} x')$ $x''$ is free for $x'$.
2. In $(\bigwedge x' F^{*\prime} x' \supset \bigwedge x'' F^{*\prime} x')$ $x''$ is not free for $x'$.
3. In $(\bigwedge x' F^{*\prime} x' \supset F^{*\prime} x')$ $f^{**\prime} x' x''$ is free for $x'$.
4. In $(\bigwedge x' F^{*\prime} x' \supset \bigwedge x'' F^{*\prime} x')$ $f^{**\prime} x' x''$ is not free for $x'$.

*Closed wffs* (or *sentences*)

A *closed* wff is a wff in which there are no free occurrences of any variable. A wff that is not closed is *open*.

*Closure*

1. If A is a wff in which the variables $v_1, \ldots, v_n$ have free occurrences, then A preceded by $\bigwedge v_1 \ldots \bigwedge v_n$ is a closure of A. [Since we do not require that the variables $v_1, \ldots, v_n$ occur free in A in that order, it follows that, e.g., both

$$\bigwedge x' \bigwedge x'' F^{**\prime} x' x''$$

and

$$\bigwedge x'' \bigwedge x' F^{**\prime} x' x''$$

are closures of

$$F^{**\prime} x' x''.]$$

2. If A is a sentence [closed wff], then A is a closure of A, and so is $\bigwedge$vA where v is any variable.

3. Any closure of a closure of A is a closure of A.

It follows that every wff has infinitely many closures.

We write '$\mathrm{A}^{c}$' for an arbitrary closure of a formula A.

*Abbreviation:* $\bigvee$ (*the existential quantifier*)

$\bigvee$vA is an abbreviation for $\sim\bigwedge v \sim A$ (v a variable, A a wff).

$\bigvee$ is called the *existential*, or *particular*, *quantifier*.

*Comments*

We assume the reader already has some familiarity with the symbolism of first order predicate logic, so we make just two remarks:

1. We want function symbols to symbolise the italicised parts of the following sentences:

(*a*) *The successor of* 0 is 1 [symbolised by a 1-place function symbol]

(*b*) *The sum of* 2 *and* 2 is 4 [symbolised by a 2-place function symbol]

(*c*) *The sum of the product of* 4 *and* 5 *and the product of* 8 *and* 10 is 100 [which could be symbolised by a 4-place function symbol: $f(a, b, c, d) = (a.b) + (c.d)$]

2. 'First order' is in contrast to 'second order' and to 'higher order'. The following account of higher order predicates and languages ignores complications connected with function symbols, function variables and propositional variables. The predicate symbols of Q are of first order. A predicate symbol that can take a predicate symbol of first order in some argument place, but never admits a predicate symbol of second or higher order in any argument place, is a second order predicate symbol; a third order predicate symbol will admit predicate symbols of second order in some argument place, but will not admit predicate symbols of higher than second order in any argument place; and so on. A first order language allows quantifiers to bind only individual variables; a second order language allows them also to bind predicate variables of first, but no higher, order; a third order language allows them also to bind predicate variables of second, but no higher, order; and so on. Or putting it semantically (which is what is intended):

First order: quantification only over individuals.

Second order: quantification over properties of[1] individuals as well as over individuals.

Third order: quantification over properties of[1] properties of[1] individuals as well as over properties of[1] individuals and over individuals.

*Q*+

Any formal language that differs from Q only in having de-

[1] or relations between.

numerably many individual constants not in Q (provided there is given an effective enumeration of these new constants) will be said to be a language Q+.

## 39 Semantics for Q (and Q+).[1] Definitions of *interpretation of Q(Q+)*, *satisfaction* of a formula by a denumerable sequence of objects, *satisfiable*, *simultaneously satisfiable*, *true for an interpretation of Q(Q+)*, *model of a formula/set of formulas of Q(Q+)*, *logically valid formula of Q(Q+)*, *semantic consequence* (for formulas of Q (Q+)), *k-validity*

Our main target in Part 3 is a proof of the completeness of a system of first order predicate logic, i.e. a proof that all logically valid formulas of Q are theorems of the system. For this we need an exact definition of *logically valid formula of Q*. We cannot just take over for this purpose the notion of *tautology* from Part 2, for there are logically valid formulas of Q that are not instances of tautological schemata (for the meaning of 'instance of a tautological schema of Q', see the definition before 40.10 below). Hence most of the complications that follow.

*Preliminary account*

An *interpretation of Q* (or *Q*+) consists in the specification of some non-empty set (called the *domain* of the interpretation) and the following assignments:

1. To each propositional symbol is assigned one or other (but not both) of the truth values truth and falsity.
2. To each individual constant is assigned some member of the domain of the interpretation.
3. To each function symbol is assigned a function with arguments and values in the domain.
4. To each predicate symbol is assigned some property or relation defined for objects in the domain.

The connectives are given their usual truth-functional meanings.[2]

[1] This kind of semantics originated with Tarski. See, e.g., Tarski (1923–38, paper VIII) (which had its beginnings *c.* 1931) or, for a simpler presentation, Tarski and Vaught (1956, pp. 84–5).

[2] But with this extension: they may stand between formulas that for a given interpretation are neither true nor false.

Quantifiers are read as referring exclusively to members of the domain of the interpretation: e.g.

$$\Lambda x'$$

is read as

For each thing, $x'$, in the domain

and $\Lambda x' F^{*\prime} x'$ is true for a given interpretation I iff every member of *the domain of I* has the property assigned by I to the predicate symbol $F^{*\prime}$.

In order to cope with wffs in which variables occur free, the full definition of *interpretation of Q* ($Q^+$) has to be rather complicated. The key notion in the definition is that of the *satisfaction* of a formula by a *denumerable sequence of objects*. Also, instead of talking about properties and relations we speak of *sets of ordered n-tuples* of objects. All this will be explained.

For convenience we recapitulate earlier material about sequences:

For our purposes we take an *ordered n-tuple* to be a *sequence of n terms*. A sequence of n terms, $\langle r_1, \ldots, r_n \rangle$, is an ordering of things, where the same thing may occur more than once in the ordering. E.g. $\langle 1, 2, 3, 1 \rangle$ is a sequence of four terms with the number 1 occurring as the first term and as the fourth term. A sequence $s$ is the same as a sequence $s'$ iff $s$ and $s'$ have exactly the same number of terms and the first term of $s$ is the same as the first term of $s'$, the second term of $s$ is the same as the second term of $s'$, and so on. So $\langle 1, 2, 3 \rangle \neq \langle 3, 2, 1 \rangle$, though $\{1, 2, 3\} = \{3, 2, 1\}$, and $\langle 1, 2, 1 \rangle \neq \langle 1, 2 \rangle$, though $\{1, 2, 1\} = \{1, 2\}$.

*Semantics for Q* ($Q^+$): *fuller account*

First we define what it is for a sequence $s$ to satisfy a formula A. We consider in turn all the forms that A may take.

*The satisfaction of a formula A by a sequence s* (*for a given interpretation I*)

Let I be an arbitrary interpretation of Q ($Q^+$), and let D be its domain. Let $s$ be an arbitrary denumerable sequence whose terms are members of D. Let A be an arbitrary formula of Q ($Q^+$).

In the definition of *satisfaction* we shall make use of the following fixed enumeration of the variables of Q ($Q^+$): $x'$ is to be the first variable in our enumeration; $x$ followed by $k$ dashes the $k$th variable in our enumeration.

1. Suppose A is a propositional symbol. Then *s* satisfies A iff I assigns the truth value truth to A.

Examples: Let I be an interpretation of Q ($Q^+$) whose domain is the set of positive integers. Let I assign to $p'$ the truth value truth, and to $p''$ the truth value falsity. For readability we shall write $p'$ as $p$, and $p''$ as $q$. So we have: I assigns T to $p$, F to $q$. Let *s* be the denumerable repeating sequence ⟨5, 10, 15, 5, 10, 15, 5, 10, 15, 5, . . .⟩. Then:

*s* satisfies $p$
*s* does not satisfy $q$

Notice that in order to know whether or not *s* satisfies a *propositional symbol*, it is not necessary to know anything about *s*, except that it is a denumerable sequence of objects from D.

2. Suppose A is of the form $\sim$B. Then *s* satisfies A iff *s* does not satisfy B.

Examples: Let I and *s* be as before. Then:

*s* does not satisfy $\sim p$
*s* satisfies $\sim q$

3. Suppose A is of the form (B $\supset$ C). Then *s* satisfies A iff either *s* does not satisfy B or *s* does satisfy C (or both). [The 'either . . . or . . .' is intended to be purely truth-functional.]

Examples: Let I and *s* be as before. Then:

*s* satisfies $q \supset p$, $q \supset q$, $q \supset \sim p$, $q \supset \sim q$
*s* satisfies $p \supset p$, $\sim p \supset p$, $\sim q \supset p$
*s* satisfies $p \supset \sim q$, $\sim p \supset \sim q$, $\sim q \supset \sim q$
*s* does not satisfy $p \supset q$, $p \supset \sim p$, $\sim q \supset q$, $\sim q \supset \sim p$

4. Suppose A is a closed atomic wff without function symbols, but not a propositional symbol. Then it is of the form $Fc_1 \ldots c_n$ where F is an $n$-place predicate symbol and $c_1, \ldots, c_n$ are individual constants (not necessarily distinct). By definition I assigns some member of D to each individual constant: let the members of D assigned to the constants $c_1, \ldots, c_n$ be $d_1, \ldots, d_n$ respectively (again, $d_1, \ldots, d_n$ are not necessarily distinct). Then *s* satisfies A iff the ordered $n$-tuple $\langle d_1, \ldots, d_n \rangle$ is a member of the set of ordered $n$-tuples assigned by I to the predicate symbol F.

Examples: Let I and $s$ be as before. Let I assign to the constant $a'$ the number 1, to $a''$ the number 2, and so on. Abbreviate $a'$ to $a_1$, $a''$ to $a_2$, and so on. Let I assign to $F^{**''}$ the relation of *being greater than* ($>$). Abbreviate $F^{**''}$ to $G$. Then:

$s$ satisfies $Ga_4a_3$ (for the number assigned to $a_4$, viz. 4, is greater than the number assigned to $a_3$, viz. 3)
$s$ does not satisfy $Ga_3a_4$
$s$ satisfies $Ga_{59}a_{58}$
$s$ does not satisfy $Ga_5a_{105}$

Again, we do not have to look at $s$.

5. Suppose A is an atomic wff of the form $\mathrm{Fv}_1 \ldots \mathrm{v}_n$ where F is an $n$-place predicate symbol and $\mathrm{v}_1, \ldots, \mathrm{v}_n$ are individual variables (not necessarily distinct). To each variable $\mathrm{v}_i$ we assign a term *in s* as follows: if $\mathrm{v}_i$ is the $k$th variable in our enumeration of all the variables of Q ($\mathrm{Q}^+$), then we assign to it the $k$th term in the sequence $s$. Let the terms in $s$ assigned to the variables $\mathrm{v}_1, \ldots, \mathrm{v}_n$ be $\mathrm{d}_1, \ldots, \mathrm{d}_n$ respectively ($\mathrm{d}_1, \ldots, \mathrm{d}_n$ are not necessarily distinct). Then $s$ satisfies A iff the ordered $n$-tuple $\langle \mathrm{d}_1, \ldots, \mathrm{d}_n \rangle$ is a member of the set of ordered $n$-tuples assigned by I to the predicate symbol F.

Examples: This time we do have to look at $s$. Let I, etc., be as before. Abbreviate $x'$ to $x_1$, $x''$ to $x_2$, etc. Then:

$s$ satisfies $Gx_3x_4$ (for the 3rd term in $s$, viz. 15, is greater than the 4th, viz. 5)
$s$ does not satisfy $Gx_2x_5$ (for the 2nd term in $s$, viz. 10, is not greater than the 5th, viz. 10)
$s$ satisfies $Gx_6x_1$
$s$ does not satisfy $Gx_1x_1$

6. Suppose A is an atomic wff of the form $\mathrm{Ft}_1 \ldots \mathrm{t}_n$ where F is an $n$-place predicate symbol and $\mathrm{t}_1, \ldots, \mathrm{t}_n$ are terms without function symbols. Each $\mathrm{t}_i$ is an individual constant or an individual variable. If $\mathrm{t}_i$ is a constant it is assigned a member (call it $\mathrm{d}_i$) of D by I. If $\mathrm{t}_i$ is a variable then there is some number $k$ such that $\mathrm{t}_i$ is the $k$th variable in our enumeration of the variables, and we assign to $\mathrm{t}_i$ the $k$th term in the sequence $s$: call this object $\mathrm{d}_i$ too. [So if $\mathrm{t}_i$ is a constant then $\mathrm{d}_i$ is the member of D assigned to the constant by I, and if $\mathrm{t}_i$ is a variable

then if $t_i$ is the $k$th variable in our enumeration then $d_i$ is the $k$th term in $s$. Every term in $s$ is a member of D.] Then $s$ satisfies A iff $\langle d_1, \ldots, d_n \rangle$ is a member of the set of ordered $n$-tuples assigned by I to the predicate symbol F.

Examples: Let I, etc., be as before. Then:

$s$ satisfies $Ga_{29}x_{10}$ (for 29 is greater than the 10th term in $s$, viz. 5)

$s$ does not satisfy $Ga_9x_{11}$ (for 9 is not greater than the 11th term in $s$, viz. 10)

7. Suppose A is of the form $\bigwedge v_k B$, where $v_k$ is the $k$th variable in our enumeration. Then $s$ satisfies A iff *every* denumerable sequence of members of D that differs from $s$ in at most the $k$th term satisfies B. I.e. $s$ satisfies A iff, no matter which member of D you substitute for the $k$th term in $s$, the resulting sequence satisfies B.

Examples: Let I, etc., be as before. Let $H$ be an abbreviation for $F^{**'''}$, to which I assigns the relation of *being greater than or equal to* ($\geqslant$). [Reminder: $s = \langle 5, 10, 15, 5, 10, 15, 5, \ldots \rangle$.] Then:

$s$ satisfies $\bigwedge x_1 Hx_1x_1$ (for every denumerable sequence of positive integers that differs from $s$ in at most the first term satisfies $Hx_1x_1$, since every positive integer is greater than *or equal to* itself)

$s$ does not satisfy $\bigwedge x_3 Hx_3x_4$ (for, though $s$ satisfies $Hx_3x_4$, not *every* denumerable sequence of positive integers that differs from $s$ in at most the 3rd term does: e.g. the sequence $\langle 5, 10, 3, 5, 10, 15, 5, 10, 15, 5, \ldots \rangle$ does not)

$s$ satisfies $\bigwedge x_{59} Hx_{59}a_1$ (for every positive integer is greater than or equal to 1)

$s$ does not satisfy $\bigwedge x_{59} Hx_{59}a_2$ (for not every positive integer is greater than or equal to 2, since 1 is not)

$s$ satisfies $\bigwedge x_{59} Hx_{59}x_{59}$

$s$ does not satisfy $\bigwedge x_{59} Hx_{59}x_1$ (not every positive integer is greater than or equal to the first term in $s$, viz. 5)

Cases 1–7 exhaust the possibilities for A, except for the cases where A contains function symbols. So now we have to put these in.

8. Suppose t is a term of the form $fc_1 \ldots c_n$, where f is an

$n$-place function symbol and $c_1, \ldots, c_n$ are individual constants. Let $f$ be the function assigned by I to f. Let $d_1, \ldots, d_n$ be the objects in D assigned by I to the constants $c_1, \ldots, c_n$. Then the object d assigned by I to t is the object in D that is the value of the function $f$ for the arguments $d_1, \ldots, d_n$: i.e. $d = f(d_1, \ldots, d_n)$.

Examples: Let I, *s*, etc., be as before. Let I assign to $f^{*\prime}$ the successor function: so '$f^{*\prime}$ . . .' means 'the successor of . . .'. Then:

$f^{*\prime}a_1$ is assigned the number 2 from D (since 2 is the successor of the number assigned to $a_1$, viz. 1)
$f^{*\prime}a_2$ is assigned the number 3
$f^{*\prime}a_{59}$ is assigned the number 60

9. Suppose t is a term of the form $fv_1 \ldots v_n$, where $v_1, \ldots, v_n$ are individual variables. Let $d_1, \ldots, d_n$ be the terms [objects] in the sequence *s* assigned to $v_1, \ldots, v_n$ in accordance with the provisions of clause 5 above. Let $f$ be as before. Then $d = f(d_1, \ldots, d_n)$.

Examples: Let I, *s*, etc., be as before. Then:

$f^{*\prime}x_1$ is assigned the number 6 (the successor of the first term in *s*, which is the number 5)
$f^{*\prime}x_2$ is assigned the number 11

10. Suppose t is a term of the form $ft_1 \ldots t_n$, where $t_1, \ldots, t_n$ are terms with function symbols followed by individual constants or variables. Let $d_1, \ldots, d_n$ be the objects in D assigned to $t_1, \ldots, t_n$ in accordance with clauses 8 and 9 above. Let $f$ be as before. Then $d = f(d_1, \ldots, d_n)$.

Examples: Let I, *s*, etc., be as before. Then:

$f^{*\prime}f^{*\prime}a_1$ is assigned the number 3 (the successor of the successor of 1)
$f^{*\prime}f^{*\prime}x_1$ is assigned the number 7 (the successor of the successor of the first term in *s*)

11. Suppose t is a term of the form $ft_1 \ldots t_n$, where $t_1, \ldots, t_n$ are arbitrary terms (including terms with function symbols). Let $d_1, \ldots, d_n$ be the objects in D assigned to $t_1, \ldots, t_n$ in accordance with clauses 8, 9 and 10 above. Let $f$ be as before. Then $d = f(d_1, \ldots, d_n)$.

Example: Let $f^{**\prime}$ be assigned the sum function: so '$f^{**\prime}xy$' means 'the sum of $x$ and $y$'. Then ($s$, etc., being as before):

$f^{**\prime}f^{*\prime}x_2f^{*\prime}f^{*\prime}a_1$ is assigned the number 14 (the sum of the number assigned to $f^{*\prime}x_2$, viz. 11, and the number assigned to $f^{*\prime}f^{*\prime}a_1$, viz. 3)

We can put this definition of satisfaction more compactly as follows. Let I be an interpretation with domain D. Let t be an arbitrary term, $s$ an arbitrary denumerable sequence of members of D. We first define a function, *, with terms of Q ($Q^+$) as arguments, and with values in D, by the following rule:

If t is a constant, then $t^*s$ is the member of D assigned by I to the constant t.
If t is the $k$th variable in our enumeration, then $t^*s$ is the $k$th term in $s$.
If t is of the form $ft_1 \ldots t_n$, where f is an $n$-place function symbol and $t_1, \ldots, t_n$ are terms, and $f$ is the function assigned by I to f, then $t^*s = f(t_1{}^*s, \ldots, t_n{}^*s)$.

Now we define *satisfaction*:

1. If A is a propositional symbol, then $s$ satisfies A iff I assigns the truth value truth to A.
2. If A is an atomic wff of the form $Ft_1 \ldots t_n$, where F is an $n$-place predicate symbol and $t_1, \ldots, t_n$ are terms, then $s$ satisfies A iff $\langle t_1{}^*s, \ldots, t_n{}^*s \rangle$ is a member of the set of ordered $n$-tuples assigned by I to F.
3. If A is of the form $\sim$B, then $s$ satisfies A iff $s$ does not satisfy B.
4. If A is of the form $(B \supset C)$, then $s$ satisfies A iff either $s$ does not satisfy B or $s$ does satisfy C.
5. If A is of the form $\Lambda v_k B$, where $v_k$ is the $k$th variable in our enumeration, then $s$ satisfies A iff every denumerable sequence of members of D that differs from $s$ in at most the $k$th term satisfies B.

[In what follows, to get the corresponding definition for $Q^+$, simply replace 'Q' by '$Q^+$' throughout.]

*Definition.* A formula A of Q is *satisfiable* iff there is some interpretation I of Q for which A is satisfied [i.e. there is an interpretation I such that A is satisfied by at least one denumerable sequence of members of the domain of I].

Examples [for simplicity we write $F$, $x$, $y$ instead of $F^{*\prime}$, $x'$, $x''$]:

1. $Fx \supset {\sim}Fy$ is satisfiable.
2. $Fx \supset {\sim}Fx$ is satisfiable.
3. ${\sim}(Fx \supset Fx)$ is not satisfiable.

*Definition.* A set Γ of formulas of Q is *simultaneously satisfiable* iff for some interpretation I of Q some sequence $s$ satisfies every member of Γ.

Examples:

1. $\{Fx, {\sim}Fy\}$ is simultaneously satisfiable.
2. $\{Fx, {\sim}Fx\}$ is not simultaneously satisfiable.

*Definition.* A wff A of Q is *true for a given interpretation I of Q* iff *every* denumerable sequence of members of the domain of I satisfies A.

*Definition.* A wff of Q is *false for a given interpretation I of Q* iff *no* denumerable sequence of members of the domain of I satisfies A.

A formula in which some variable has a free occurrence *may* be neither true nor false for a given interpretation. E.g. $Fx \supset {\sim}Fy$ is neither true nor false for an interpretation I in which D is the set of natural numbers and '$F$' means 'is even'. Cf. 40.6 and 40.7 on pp. 152, 153.

N.B. '*To satisfy*' *does not mean* '*to make true*'. A sequence $s$ may *satisfy* a formula A, for an interpretation I, without A being *true* for I. For example: Let I, $s$, $G$, $x_1$, $x_2$ be as in the examples on pp. 143–4 above. Then $s$ *satisfies* $Gx_2x_1$ (for 10 is greater than 5), but $Gx_2x_1$ is *not* true for I (for not *every* sequence of members of the domain of I satisfies $Gx_2x_1$: the sequence $\langle 2, 1, 1, 1, 1, 1, 1, 1, \ldots \rangle$, for example, does not satisfy it).

*Definition.* An interpretation I of Q is a *model of a formula A of Q* iff A is true for I.

*Definition.* An interpretation I of Q is a *model of a set Γ of formulas of Q* iff every formula in Γ is true for I.

*Definition.* A *formal system has a model* iff the set of all its theorems has a model.

*Definition.* A formula A of Q is a *logically valid formula of Q* [$\models_Q$ A] iff A is true for *every* interpretation of Q.

[Remember that every interpretation of Q must, by definition, have a non-empty domain.]

*Definition.* A formula B of Q is a *semantic consequence of a formula A of Q* [A $\models_Q$ B] iff for every interpretation of Q every sequence that satisfies A also satisfies B: i.e. there is no sequence that satisfies A and does not also satisfy B.

If no sequence satisfies A, then any wff of Q you like to take is a semantic consequence of A.

*Definition.* A formula A of Q is a *semantic consequence of a set Γ of formulas of Q* [Γ $\models_Q$ A] iff for every interpretation of Q every sequence that satisfies every member of Γ also satisfies A: i.e. there is no sequence that satisfies every member of Γ and does not also satisfy A.

*Note.* It is worth comparing these definitions with the corresponding definitions for P: e.g. 'A formula B of P is a semantic consequence of a formula A of P iff there is no interpretation of P for which A is true and B is false'. It is *not* a sufficient condition for an arbitrary formula B of Q being a semantic consequence of a formula A of Q that there is no interpretation of Q for which A is true and B is false. The requirement for being a semantic consequence for formulas of Q is stronger. For example, there is no interpretation of Q for which *Fx* is true and $\bigwedge xFx$ is false, but $\bigwedge xFx$ is not a semantic consequence of *Fx*, for it is not true that for every interpretation of Q every sequence that satisfies *Fx* also satisfies $\bigwedge xFx$.

*Definition.* A formula of Q is *k-valid* iff it is true for every interpretation of Q that has a domain of exactly *k* members.

*The empty set*, ∅

By convention every denumerable sequence satisfies the empty set, ∅. So:

39.1 $\emptyset \models_Q A$ *iff* $\models_Q A$ [Similarly for $Q^+$]

Cf. 19.1.

*Final note to* §39

Basic to §39 are:

I. The difference between

(1) *s satisfies A*,

(2) *A is true for I*, and

(3) *A is logically valid.*

II. The notion of *semantic consequence for Q*.

The reader is advised not to go on until he is reasonably clear about I and II.

EXERCISES

1. Let $I_1$ be an interpretation of Q whose domain is the set of positive integers {1, 2, 3, . . .}. $I_1$ assigns to the predicate symbol $G$ [i.e. $F^{**''}$] the relation $\leqslant$. Let $s$ be the denumerable sequence $\langle 2, 5, 8, 1, 7, 3, 3, 3, \ldots, 3, \ldots \rangle$. We abbreviate $x', x'', x''', \ldots$ to $x_1, x_2, x_3, \ldots$ and $F^{*'}$ to $F$.

(i) Does $s$ satisfy the formula $Gx_2x_3$ (for the interpretation $I_1$)?

(ii) Does $s$ satisfy the formula $Gx_3x_4$ (for the interpretation $I_1$)?

(iii) Is $Gx_2x_3$ true for $I_1$?

(iv) Is $Gx_2x_2$ true for $I_1$?

(v) Is $I_1$ a model of the set of formulas $\{Gx_{49}x_{107}, \bigwedge x_2 \bigwedge x_3 Gx_2x_3\}$?

(vi) Is $I_1$ a model of the set of formulas $\{Gx_5x_5, \bigwedge x_{72} Gx_{72}x_{72}\}$?

(vii) Is $Gx_5x_5$ logically valid?

(viii) Is $Gx_2x_1$ satisfiable?

(ix) Is $Fx_{13}$ a semantic consequence of $\bigwedge x_2 Fx_2$?

(x) Is $Fx_{13}$ a semantic consequence of $\sim \bigwedge x_2 Fx_2$?

(xi) Is $Fx_{13}$ a semantic consequence of $\sim(\bigwedge x_2 Fx_2 \supset \bigwedge x_2 Fx_2)$?

2. Let $I_2$ be an interpretation of Q whose domain is the set of even numbers {2, 4, 6, 8, . . .}. $I_2$ assigns to the predicate symbol $G$ the relation $\leqslant$.

(i) Specify a denumerable sequence of objects that satisfies the formula $Gx_4x_5$ for the interpretation $I_2$.

(ii) Specify a denumerable sequence of objects that does not satisfy the formula $Gx_4x_5$ for the interpretation $I_2$.

(iii) Exhibit a formula that is true for $I_2$.

(iv) Exhibit a formula that is not true for $I_2$.

(v) Exhibit a set of formulas for which $I_2$ is a model.
(vi) Exhibit a set of formulas for which $I_2$ is not a model.
(vii) Exhibit a formula that is logically valid.
(viii) Exhibit a formula that is not logically valid.
(ix) Exhibit a formula that is satisfiable.
(x) Exhibit a formula that is not satisfiable.

ANSWERS

1. (i) Yes.
(ii) No.
(iii) No. A formula is *true* for $I_1$ iff it is satisfied by *every* denumerable sequence of positive integers. The sequence $\langle 2, 8, 2, 2, 2, \ldots \rangle$ (e.g.) does not satisfy $Gx_2x_3$.
(iv) Yes: but not merely because $5 \leqslant 5$, but because, for *each* positive integer $n$, $n \leqslant n$.
(v) No. *Every* denumerable sequence of positive integers has to satisfy both formulas simultaneously if $I_1$ is to be a model of this set. One sequence (among many) that does not satisfy $Gx_{49}x_{107}$ is a sequence whose 49th term is 2 and whose 107th term is 1. So $I_1$ is not a model of the set.
(vi) Yes.
(vii) No. It is true for $I_1$, but not true for *every* interpretation of Q; e.g. it is not true for an interpretation in which $G$ is assigned the relation $>$.
(viii) Yes.
(ix) Yes.
(x) No.
(xi) Yes. No sequence satisfies $\sim(\Lambda x_2Fx_2 \supset \Lambda x_2Fx_2)$.
2. (i) E.g. the sequence $\langle 2, 4, 6, 8, 10, \ldots \rangle$ or the sequence $\langle 2, 4, 6, 8, 8, 10, 12, \ldots \rangle$ or the sequence $\langle 8, 6, 4, 2, 4, 6, 8, 6, 4, 2, 4, \ldots \rangle$.
(ii) E.g. the sequence $\langle 2, 4, 6, 8, 6, 4, 2, 4, 6, \ldots \rangle$.
(iii) E.g. $\Lambda x_5Gx_5x_5$.
(iv) E.g. $Gx_1x_2$.
(v) E.g. $\{Gx_{17}x_{17}, \Lambda x_9Gx_9x_9\}$ or $\{Gx_{17}x_{17}, \Lambda x_9Gx_9x_9, p' \supset p'\}$.
(vi) E.g. $\{Gx_{17}x_{17}, \Lambda x_9Gx_9x_9, Gx_1x_2\}$ or $\{Gx_{17}x_{17}, \Lambda x_9Gx_9x_9, \sim(p' \supset p')\}$.
(vii) E.g. $Fx_1 \supset Fx_1$ or $Fx_1 \supset \Lambda x_2Fx_1$ [cf. 40.13 below].
(viii) E.g. $Fx_1 \supset \Lambda x_1Fx_1$ [e.g. take an interpretation $I_3$ with

domain {1, 2} that assigns the property of *being an even number* to $F$. Then the sequence ⟨2, 2, 2, 2, 2, . . .⟩ satisfies $Fx_1$ but does not satisfy $\Lambda x_1 Fx_1$].

(ix) E.g. $Fx_1 \supset Fx_2$.

(x) E.g. $\sim(Fx_1 \supset Fx_1)$.

## 40 Some model-theoretic metatheorems for Q (and Q+)

[This section assembles various results that will be appealed to later. It could be skipped at a first reading. Alternatively the proofs of the metatheorems could be skipped at a first reading.]

We develop only as much of the model theory of Q (Q+) as we need for later proofs.

The following are more or less immediate consequences of the definitions in §39:

40.1 *If A is logically valid, then* $\sim A$ *is not satisfiable*

40.2 *Modus Ponens for* $\supset$ *preserves satisfaction-by-s [i.e. if a sequence s satisfies A and also* $A \supset B$*, then it also satisfies B]*

40.3 *Modus Ponens for* $\supset$ *preserves truth-for-I [i.e. if A and* $A \supset B$ *are both true for an interpretation I, then B is also true for I]*

40.4 *Modus Ponens for* $\supset$ *preserves logical validity [i.e. if A and* $A \supset B$ *are both logically valid, then so is B. Or: If* $\models_Q A$ *and* $\models_Q A \supset B$*, then* $\models_Q B$*]*

[The reference to Modus Ponens, and therefore proof theory, in the statements of 40.2–40.4 is inessential. It was put in simply to indicate a later use for those metatheorems.]

40.5 *A is false for a given interpretation I iff* $\sim A$ *is true for I; and A is true for I iff* $\sim A$ *is false for I*

40.6 *A is true for I iff* $\Lambda vA$ *is true for I for any arbitrary variable v*[1]

[1] If A is false for I, then ΛvA is false for I. But the converse is not true; if A contains a free variable, ΛvA may be false for I while A is neither true nor false for I.

40.7 *A is true for I iff any arbitrary closure of A is true for I*

40.8 *A is logically valid iff $A^c$ is* [$A^c$ is an arbitrary closure of A]

Less immediate are the following:

40.9 *$\bigvee$vA is satisfiable for an interpretation I iff A is satisfiable for the same interpretation*

*Proof.* Let I be an interpretation with domain D, and *s* and *s′* denumerable sequences of members of D.

1. Let *s* be any sequence that satisfies $\bigvee$vA, i.e. $\sim\bigwedge$v$\sim$A. Then *s* does not satisfy $\bigwedge$v$\sim$A. So not every denumerable sequence of members of D satisfies $\sim$A. Let *s′* be a sequence that does not satisfy $\sim$A. Then *s′* satisfies A. So if $\bigvee$vA is satisfiable for I, then A is satisfiable for I.

2. Suppose no denumerable sequence of members of D satisfies $\bigvee$vA, i.e. $\sim\bigwedge$v$\sim$A. Then every sequence of members of D satisfies $\bigwedge$v$\sim$A. So every sequence satisfies $\sim$A for I. So no sequence satisfies A for I. So if some sequence satisfies A for I, then some sequence satisfies $\bigvee$vA for I.

*Definition.* If A is a tautology of P whose only propositional symbols are $P_1, \ldots, P_n$, then the result of substituting wffs $Q_1, \ldots, Q_n$ of Q for $P_1, \ldots, P_n$ respectively in A is an *instance of a tautological schema of Q*.

Similarly for $Q^+$.

40.10 *Every instance of a tautological schema of Q ($Q^+$) is logically valid*

*Outline of proof.* Let X be an arbitrary tautological schema of Q. Let *s* be an arbitrary denumerable sequence of members of the domain of an arbitrary interpretation of Q ($Q^+$). It is a consequence of the definition of satisfaction that, for any wff A of Q ($Q^+$), either *s* satisfies A or *s* does not satisfy A. We make out a table for X that is like a truth table except that instead of 'T' and 'F' we write 'Yes' (meaning '*s* satisfies') and 'No' (meaning '*s* does not satisfy'). Then, by clauses 3 and 4 of the short definition of *satisfaction* on p. 147, X will get 'Yes' for every assignment of 'Yes'es and No'es' to its schematic letters. But *s* was an arbitrary sequence. So every sequence satisfies any instance of X.

Example: Let X be the schema $A \supset (\sim B \supset \sim(A \supset B))$

| A | B | A | $\supset$ | $(\sim$ | B | $\supset$ | $\sim$ | (A | $\supset$ | B)) |
|---|---|---|---|---|---|---|---|---|---|---|
| Yes | Yes | Yes | Yes | No | Yes | Yes | No | Yes | Yes | Yes |
| No | Yes | No | Yes | No | Yes | Yes | No | No | Yes | Yes |
| Yes | No | Yes | Yes | Yes | No | Yes | Yes | Yes | No | No |
| No | No | No | Yes | Yes | No | No | No | No | Yes | No |
| | | | * | | | | | | | |

40.11 $\Lambda v_k(A \supset B) \supset (\Lambda v_k A \supset \Lambda v_k B)$ *is logically valid, for arbitrary wffs A and B and an arbitrary variable* $v_k$

*Proof.* Suppose not. Then for some interpretation there is a sequence *s* that satisfies [an instance of] $\Lambda v_k(A \supset B)$ and [the corresponding instance of] $\Lambda v_k A$ but does not satisfy [the corresponding instance of] $\Lambda v_k B$. Now *s* satisfies $\Lambda v_k(A \supset B)$ and $\Lambda v_k A$ iff every sequence that differs from *s* in at most the *k*th term satisfies $A \supset B$ and A. But by 40.2 every such sequence also satisfies B, and therefore *s* satisfies $\Lambda v_k B$. But this contradicts our initial supposition.

Notation:

Let I be an interpretation, D its domain, *s* a denumerable sequence of members of D.

$s(d/k)$ is the sequence that results from replacing the *k*th term in the sequence *s* by the object d.

$t^*s$ is the member of D assigned by I to the term t for the sequence *s*, as in §39: i.e.:

If t is an individual constant c, then $t^*s$ is the member of D assigned by I to c.

If t is $v_k$, then $t^*s$ is the *k*th term in *s*.

If t is an *n*-place function symbol f followed by *n* terms, $t_1, \ldots, t_n$, and *f* is the function assigned by I to f, then $t^*s = [ft_1 \ldots t_n]^*s = f(t_1{}^*s, \ldots, t_n{}^*s)$.

40.12 *Let I be an interpretation with domain D. Let A be an arbitrary wff. Let s and s′ be two sequences such that, for each free variable v in A, if v is the kth variable in the fixed enumeration of the variables, then s and s′ have the same member of D for their kth terms. Then s satisfies A iff s′ does*

*Proof.*[1] By induction on the number, *n*, of connectives and quantifiers in A.

[1] Could be skipped at a first reading.

*Basis*: $n = 0$

Then A is atomic. 2 cases:

1. A is a propositional symbol
2. A is of the form $\mathrm{F}\mathrm{t}_1 \ldots \mathrm{t}_m$, where F is an $m$-place predicate symbol and $\mathrm{t}_1, \ldots, \mathrm{t}_m$ are terms

*Case* 1. Obvious.

*Case* 2. Let I assign the relation R to F (If F is a 1-place predicate symbol, then R is a property. But for brevity we speak of R as a relation throughout). $s$ satisfies A iff $\mathrm{t}_1{}^*s, \ldots, \mathrm{t}_m{}^*s$ in that order stand in the relation R, and $s'$ satisfies A iff $\mathrm{t}_1{}^*s', \ldots, \mathrm{t}_m{}^*s'$ stand in the relation R. We shall prove that, for each $\mathrm{t}_i$, $\mathrm{t}_i{}^*s = \mathrm{t}_i{}^*s'$.

(i) If $\mathrm{t}_i$ is a constant, then $\mathrm{t}_i{}^*s = \mathrm{t}_i{}^*s'$.

(ii) If $\mathrm{t}_i$ is a variable $\mathrm{v}_k$, then since A is atomic $\mathrm{t}_i$ is free, and therefore by the hypothesis of the theorem the $k$th term of $s$ = the $k$th term of $s'$; i.e. $\mathrm{t}_i{}^*s = \mathrm{t}_i{}^*s'$.

(iii) Suppose $\mathrm{t}_i$ is $\mathrm{f}\mathrm{t}_{j_1} \ldots \mathrm{t}_{j_r}$ (i.e. an $r$-place function symbol followed by $r$ terms). If each of $\mathrm{t}_{j_1}, \ldots, \mathrm{t}_j$ is a constant or free variable, then $\mathrm{t}_{j_i}{}^*s = \mathrm{t}_{j_i}{}^*s'$, by (i) and (ii) above. If any function symbols occur in $\mathrm{t}_{j_i}$, then, since the functions assigned by I to any function symbols are the same for $s$ and $s'$, their values for the same arguments will be the same. This holds too for the function assigned by I to f. So in this case too $\mathrm{t}_i{}^*s = \mathrm{t}_i{}^*s'$.

*Induction Step*

Assume the theorem holds for all A with fewer than $q$ connectives and quantifiers. To prove it holds for all A with exactly $q$ connectives and quantifiers. 3 cases:

1. A is $\sim$B
2. A is $\mathrm{B} \supset \mathrm{C}$
3. A is $\Lambda \mathrm{v}_p \mathrm{B}$ [$\mathrm{v}_p$ is a variable. We do not use '$k$' as a subscript here because in the proof for Case 3 below $k$ and $p$ may be distinct.]

*Case 1.* A is $\sim$B. By the induction hypothesis $s$ satisfies B iff $s'$ does. So $s$ does not satisfy A iff $s'$ does not satisfy A.

*Case 2.* A is $\mathrm{B} \supset \mathrm{C}$. Similarly.

*Case 3.* A is $\Lambda \mathrm{v}_p \mathrm{B}$. Suppose $s$ satisfies A. Then $s(\mathrm{d}_i/\mathrm{p})$ satisfies B for every $\mathrm{d}_i$ in the domain D. For each free variable $\mathrm{v}_k$ in A $s$

and $s'$ have their $k$th term in common, *ex hypothesi*. So for each $d_i$ and each free variable $v_k$ in A $s(d_i/p)$ and $s'(d_i/p)$ have their $k$th term in common. (Intuitively: If the $p$th term is the $k$th term for some free variable $v_k$ in A, then the $p$th term is the same in $s(d_i/p)$ and $s'(d_i/p)$, viz. $d_i$. If it is not the $k$th term for a free $v_k$, then still $s(d_i/p)$ and $s'(d_i/p)$ have their $k$th term in common, since $s$ and $s'$ do.) So by the induction hypothesis $s'(d_i/p)$ satisfies B for every $d_i$ in D. So $s'(d_i/p)$ satisfies A. By an exactly similar argument, if $s'$ satisfies A then $s$ does.

40.13 *If $v_k$ does not occur free in A, then $A \supset \Lambda v_k A$ is logically valid (A an arbitrary wff)*

*Proof.* Suppose $s$ satisfies A. Let $s'$ be any sequence that differs from $s$ at most in the $k$th term. Then the hypothesis of 40.12 applies to $s$ and $s'$, and so $s'$ satisfies A. But $s'$ was an *arbitrary* sequence that differs from $s$ in at most the $k$th term. So every sequence that differs from $s$ in at most the $k$th term satisfies A. So $s$ satisfies $\Lambda v_k A$.

40.14[1] *Let t and u be terms. Let t' be the result of replacing each occurrence of $v_k$ in t by u. Let s be a sequence, and let $u^*s = d$: i.e. let the member of D assigned by I to u for the sequence s be d. Let s' be $s(d/k)$: i.e. let s' be the sequence that results from substituting d for the kth term of s. Then $t'^*s = t^*s'$: i.e. the member of D assigned by I to t' for the sequence s is the same as the member of D assigned by I to t for the sequence s'.*

Example: Let t be $v_k$. Let u be $v_j$. Then t' is $v_j$. Then d is the $j$th term in $s$; and $s'$ differs from $s$ at most in having d for its $k$th term. Then the theorem says that the member of D assigned by I to t' for $s$ (in this case the $j$th term in $s$, i.e. d) is the same as the member of D assigned by I to t for $s'$ (in this case the $k$th term in $s'$, i.e. d).

*Proof.* The proof is by induction on the length of t, where the length of t is given by the number of occurrences of individual symbols (i.e. individual constants or individual variables) and function symbols in t.

[1] Could be skipped at a first reading.

*Basis*: $n = 1$

Then t is either an individual constant or an individual variable. 3 cases:

1. t is a constant
2. t is a variable $v_j$ and $j \neq k$
3. t is a variable $v_j$ and $j = k$

It is a mechanical task to verify that the theorem holds for each of these cases.

*Induction Step*

Assume the theorem holds for all terms of length less than $q$. To prove it holds for all terms of length $q$.

In view of the Basis we need only consider cases where $n > 1$. Then t must be of the form $ft_1 \ldots t_m$, where f is an $m$-place function symbol and $t_1, \ldots, t_m$ are terms of length less than $q$ (so the induction hypothesis can be applied to them). Let the function assigned by I to f be $f$. t′ is $ft_1' \ldots t_m'$. Then $t'^*s = [ft_1' \ldots t_m']^*s = f(t_1'^*s, \ldots, t_m'^*s) =$ [by the induction hypothesis] $f(t_1^*s', \ldots, t_m^*s') = [ft_1 \ldots t_m]^*s' = t^*s'$.

*Reminders* (§38)

If A is a wff, t a term, v a variable, then At/v is the wff got from A by substituting t for all free occurrences of v in A.

*t is free for v in A* if (1) if t is a variable, then t occurs free in At/v wherever v occurs free in A, and (2) if t is a term in which any variables occur, then wherever t is substituted for free occurrences of v in A all occurrences of variables in t remain free. If t is a closed term, then t is free for v in any A.

40.15[1] *Let A be a wff, $v_k$ a variable, t a term that is free for $v_k$ in A. Let s be a sequence, and let s′ be the sequence that results from replacing the kth term of s by $t^*s$ (i.e. the member of D assigned by I to the term t for the sequence s): i.e. $s' = s(t^*s/k)$. Then s satisfies $At/v_k$ iff s′ satisfies A*

*Proof.* By induction on the number of quantifiers and connectives in A.

*Basis*: $n = 0$

Then A is either a propositional symbol or of the form $Ft_1 \ldots t_m$. Clearly the theorem holds if A is a propositional symbol. Let

[1] Could be skipped at a first reading.

$t_i'$ be the result of replacing each occurrence of $v_k$ in $t_i$ by t. Then $At/v_k$ is $Ft_1' \ldots t_m'$. Let R be the relation assigned to F by I. Then $s$ satisfies $At/v_k$ iff $t_1'^*s', \ldots, t_m'^*s$ stand in the relation R. $s'$ satisfies A iff $t_1^*s', \ldots, t_m^*s'$ stand in the relation R. By 40.14 $t_i'^*s = t_i^*s'$ for $1 \leqslant i \leqslant m$. So $s$ satisfies $At/v_k$ iff $s'$ satisfies A.

*Induction Step*

Assume the theorem holds for all wffs with fewer than $q$ connectives and quantifiers. To prove it holds for all wffs with $q$ connectives and quantifiers. 3 cases:

1. A is $\sim$B
2. A is B$\supset$C
3. A is $\bigwedge v_j B$

*Case 1.* A is $\sim$B. Then $At/v_k$ is $\sim Bt/v_k$. $s$ satisfies $\sim Bt/v_k$ iff it does not satisfy $Bt/v_k$. t is free for $v_k$ in A, so it is free for $v_k$ in B also. So by the induction hypothesis $s$ does not satisfy $Bt/v_k$ iff $s'$ does not satisfy B. But $s'$ does not satisfy B iff $s'$ satisfies A. So $s$ satisfies $At/v_k$ iff $s'$ satisfies A.

*Case 2.* A is B$\supset$C. Obvious.

*Case 3.* A is $\bigwedge v_j B$. Then $At/v_k$ is $\bigwedge v_j Bt/v_k$, and t is free for $v_k$ in B.

(i) Suppose $v_k$ is free in A and $j \neq k$. $s$ satisfies $At/v_k$ iff $s$(d/j) satisfies $Bt/v_k$ for every member d of D. By the induction hypothesis $s$(d/j) satisfies $Bt/v_k$ iff $s$(d/j)$'$ satisfies B. Since t is free for $v_k$ in B and A is $\bigwedge v_j B$, $v_j$ does not occur in t. So $t^*s$ does not depend on what the $j$th term in $s$ is, for any $s$. So, for any d, $s$(d/j)$'$, which is the sequence that results from replacing the $k$th term in $s$(d/j) by $t^*s$, is $s'$(d/j), which is the sequence that results from first replacing the $k$th term in $s$ by $t^*s$ and then replacing the $j$th term in the resulting sequence by d. $s'$ satisfies A iff $s'$(d/j) satisfies B for every d. So we have:

$s$ satisfies $At/v_k$ iff $s$(d/j) satisfies $Bt/v_k$ for every d.
$s$(d/j) satisfies $Bt/v_k$ iff $s$(d/j)$'$ satisfies B.
For any d, $s$(d/j)$'$ = $s'$(d/j).
$s'$ satisfies A iff $s'$(d/j) satisfies B for every d.

So $s$ satisfies $At/v_k$ iff $s'$ satisfies A.

(ii) Suppose $v_k$ is not free in A. Then $At/v_k$ is A. Since $s'$ differs from $s$ at most in the $k$th term, and $v_k$ is not a free variable of A, we can apply 40.12 and get:

$s$ satisfies A iff $s'$ satisfies A
i.e. $s$ satisfies $At/v_k$ iff $s'$ satisfies A.

(iii) Suppose $j = k$. Then $v_k$ is not free in A. Then as for (ii) above.

40.16 *$\Lambda v_k A \supset At/v_k$ is logically valid if t is free for $v_k$ in A*

*Proof.* Suppose $s$ satisfies $\Lambda v_k A$. Then $s(d/k)$ satisfies A for every d. So $s(t^*s/k)$ satisfies A. So by 40.15 $s$ satisfies $At/v_k$.

40.17 *If A is a closed wff, then exactly one of A and ~A is true and exactly one false*

*Proof.* Since A has no free variables, the hypothesis of 40.12 is satisfied by *any* two sequences. So by 40.12 a sequence satisfies A iff *every* sequence does. So either every sequence satisfies A or none does. So either A is true for I or A is false for I. If A is true for I then no sequence satisfies ~A, i.e. ~A is false for I. If A is false for I, then ~A is true for I.

40.18 *If A and B are closed wffs, then $A \supset B$ is true for I iff A is false for I or B is true for I*

*Proof*

(*a*) Suppose $A \supset B$ is true for I. Then an arbitrary sequence $s$ satisfies $A \supset B$. So $s$ does not satisfy A or $s$ does satisfy B. If $s$ does not satisfy A, then A is not true for I, and so by 40.17 A is false for I. If $s$ satisfies B, then B is not false for I, and so by 40.17 B is true for I.

(*b*) Suppose A is false for I or B is true for I. (i) If A is false for I, then no sequence satisfies A and so every sequence satisfies $A \supset B$. (ii) If B is true for I, then every sequence satisfies B, and so every sequence satisfies $A \supset B$. So in either case $A \supset B$ is true for I.

40.19 *If A and B are closed wffs, then $A \supset B$ is false for I iff A is true for I and B is false for I*

*Proof.* If A and B are closed wffs so is $A \supset B$. By 40.17 $A \supset B$ is false for I iff $A \supset B$ is not true for I, and by 40.18 $A \supset B$ is not true for I iff A is not false for I and B is not true for I, i.e. (since A and B are closed wffs) iff A is true for I and B is false for I.

40.20[1] *If a formula A with exactly one free variable $v_k$ is true for I, then each formula that results from substituting a closed term for the free occurrences of the variable is true for I*

*Proof.* Suppose A is true for I. Then by 40.6 its closure $\Lambda v_k A$ is true for I. By 40.16 $\Lambda v_k A \supset At/v_k$ is logically valid if t is a closed term, so if $\Lambda v_k A$ is true, then $At/v_k$ is true if t is a closed term. So $At/v_k$ is true for I for every closed term t.

40.21[1] *Let I be an interpretation with domain D. Let A be a wff with exactly one free variable, $v_k$. If each member of D is assigned by I to some closed term or other, and $At/v_k$ is true for I for each closed term t, then $\Lambda v_k A$ is true for I*

*Proof.* Suppose $At/v_k$ is true for I for each closed term t. Let c be an arbitrary closed term. Then $Ac/v_k$ is true for I. Let *s* be an arbitrary sequence. Then *s* satisfies $Ac/v_k$. Let *s'* be *s*(c**s*/k). Then by 40.15 *s'* satisfies A. But *s* was an arbitrary sequence and c an arbitrary closed term. So, since *ex hypothesi* each member of D is assigned to some closed term, *s'* may be *any* arbitrary sequence. So A is satisfied by any arbitrary sequence. So $\Lambda v_k A$ is true for I.

For the convenience of the reader we repeat here the definition of *k-validity*:

*Definition.* A formula of Q is *k-valid* iff it is true for every interpretation of Q that has a domain of exactly *k* members.

40.22[2] *There is an effective method for telling, given an arbitrary formula A of Q and an arbitrary positive integer k, whether or not A is k-valid*

*Informal proof.* For a particular positive integer *k* there are infinitely many different interpretations of Q that have domains with exactly *k* members. But in order to determine the *k*-validity of a formula it is not necessary to consider each of the infinitely many possible interpretations of (say) a predicate symbol. There are only finitely many *relevantly* different possibilities that have to be considered (we consider classes of interpretations rather than particular interpretations). To illustrate:

[1] Used in the completeness proof, but could be skipped at a first reading.
[2] Used in the proof of the decidability of a system of monadic predicate logic, §50.

*Illustration 1*

Let the formula be $F^{*\prime}x^{\prime\prime\prime}$. Let $k = 2$. Let I be an interpretation with a domain D of exactly two members, say $d_1$ and $d_2$. Then, by 40.12, in order to tell whether or not an arbitrary denumerable sequence $s$ of members of D satisfies $F^{*\prime}x^{\prime\prime\prime}$, we need only consider whether or not the third term of the sequence $s$ does or does not possess the property assigned by I to the predicate symbol $F^{*\prime}$. The other terms in the sequence do not matter. So, since D has only two members, $d_1$ and $d_2$, there are just two relevantly different cases that we have to consider, viz. the case where $s$ has $d_1$ for its third term and the case where $s$ has $d_2$ for its third term. Let $s_1$ be an arbitrary denumerable sequence of members of D whose third term is $d_1$, and let $s_2$ be an arbitrary sequence whose third term is $d_2$. Abbreviate $F^{*\prime}x^{\prime\prime\prime}$ to $Fx_3$. Then we get the following four possibilities:

Possibilities:

1 $s_1$ satisfies $Fx_3$
   $s_2$ satisfies $Fx_3$

2 $s_1$ does not satisfy $Fx_3$
   $s_2$ satisfies $Fx_3$

3 $s_1$ satisfies $Fx_3$
   $s_2$ does not satisfy $Fx_3$

4 $s_1$ does not satisfy $Fx_3$
   $s_2$ does not satisfy $Fx_3$

*Illustration 2*

Let the formula be $F^{**\prime}x^{\prime}x^{\prime\prime}$. Let $k$ again be 2.

This time there are four relevantly different kinds of sequence.

Let $s_1$ be an arbitrary sequence beginning $\langle d_1, d_1, \ldots \rangle$.

Let $s_2$ be an arbitrary sequence beginning $\langle d_1, d_2, \ldots \rangle$.

Let $s_3$ be an arbitrary sequence beginning $\langle d_2, d_2, \ldots \rangle$.

Let $s_4$ be an arbitrary sequence beginning $\langle d_2, d_1, \ldots \rangle$.

Abbreviate $F^{**\prime}x^{\prime}x^{\prime\prime}$ to $A$. Then we get 16 relevantly different possibilities:

| | Possibilities | | | | | | |
|---|---|---|---|---|---|---|---|
| | 1 | 2 | 3 | 4 | ....... | 15 | 16 |
| $s_1$ satisfies $A$? | Yes | No | Yes | No | ....... | Yes | No |
| $s_2$ satisfies $A$? | Yes | Yes | No | No | ....... | No | No |
| $s_3$ satisfies $A$? | Yes | Yes | Yes | Yes | ....... | No | No |
| $s_4$ satisfies $A$? | Yes | Yes | Yes | Yes | ....... | No | No |

*Illustration 3*

Let the formula be $F^{**\prime}x'x''$, abbreviated to $A$. Let $k = 3$. Let I be an arbitrary interpretation with a domain of three members, $d_1$, $d_2$ and $d_3$.

This time we get nine relevantly different kinds of sequence. Let $s_1$ be an arbitrary sequence beginning $\langle d_1, d_1, \ldots \rangle$

| | |
|---|---|
| $s_2$ | $\langle d_1, d_2, \ldots \rangle$ |
| $s_3$ | $\langle d_1, d_3, \ldots \rangle$ |
| $s_4$ | $\langle d_2, d_1, \ldots \rangle$ |
| $s_5$ | $\langle d_2, d_2, \ldots \rangle$ |
| $s_6$ | $\langle d_2, d_3, \ldots \rangle$ |
| $s_7$ | $\langle d_3, d_1, \ldots \rangle$ |
| $s_8$ | $\langle d_3, d_2, \ldots \rangle$ |
| $s_9$ | $\langle d_3, d_3, \ldots \rangle$ |

We get 512 [ $= 2^{(3^2)}$] possibilities:

| | Possibilities | | | | | | | |
|---|---|---|---|---|---|---|---|---|
| | 1 | 2 | 3 | ... | ... | ... | 511 | 512 |
| $s_1$ satisfies $A$? | Yes | No | Yes | ... | ... | ... | Yes | No |
| $s_2$ satisfies $A$? | Yes | Yes | No | ... | ... | ... | No | No |
| $s_3$ satisfies $A$? | Yes | Yes | Yes | ... | ... | ... | No | No |
| ... | ... | ... | ... | ... | ... | ... | ... | ... |
| ... | ... | ... | ... | ... | ... | ... | ... | ... |
| $s_9$ satisfies $A$? | Yes | Yes | Yes | ... | ... | ... | No | No |

*Illustration 4*

Let the formula be $F^{**\prime}a'a''$, abbreviated to $B$. Let $k = 2$. Let I be an arbitrary interpretation with a domain D of two members, $d_1$ and $d_2$. There are four possible assignments to $a'$, $a''$, viz.

1. I assigns $d_1$ to both $a'$ and $a''$
2. I assigns $d_1$ to $a'$ and $d_2$ to $a''$
3. I assigns $d_2$ to both $a'$ and $a''$
4. I assigns $d_2$ to $a'$ and $d_1$ to $a''$

For each of these assignments an arbitrary sequence $s$ may either satisfy or not satisfy $B$. This gives 16 possibilities:

| I assigns to: | | | Possibilities | | | | | | |
|---|---|---|---|---|---|---|---|---|---|
| $a'$ | $a''$ | | 1 | 2 | 3 ... | ... | 15 | 16 |
| $d_1$ | $d_1$ | $s$ satisfies $B$? | Yes | No | Yes ... | ... | Yes | No |
| $d_1$ | $d_2$ | $s$ satisfies $B$? | Yes | Yes | No ... | ... | No | No |
| $d_2$ | $d_2$ | $s$ satisfies $B$? | Yes | Yes | Yes ... | ... | No | No |
| $d_2$ | $d_1$ | $s$ satisfies $B$? | Yes | Yes | Yes ... | ... | No | No |

To determine whether or not a particular formula is $k$-valid for a particular positive integer $k$, all we have to do is run through the *finitely* many different possibilities that arise from permuting the relevantly different possibilities for the atomic parts of the formula. Since in any given case the number of possibilities is finite, it is always possible in principle to run through them all, though, as can be seen, if $k$ is large or the formula contains $n$-place predicate symbols with $n$ large, the number of possibilities to be considered may be very large.

*Example 1*: Is $F^{*\prime}x' \supset \bigwedge x' F^{*\prime}x'$ 1-valid?

Let I be an arbitrary interpretation with a domain D with exactly one member, say d. Then there is only one denumerable sequence $s$ of members of D, viz. the sequence ⟨d, d, d, . . .⟩. There are just two relevant possibilities for $F^{*\prime}x'$: $s$ either does or does not satisfy it. If $s$ does satisfy it, then $s$ also satisfies $\bigwedge x' F^{*\prime}x'$ (since $s$ is the only denumerable sequence of members of D). If $s$ does not satisfy $F^{*\prime}x'$, then $s$ does not satisfy $\bigwedge x' F^{*\prime}x'$. So we get the following table, where 'Yes' means '$s$ satisfies' and 'No' means '$s$ does not satisfy':

| $F^{*\prime}x'$ | $\bigwedge x' F^{*\prime}x'$ | $F^{*\prime}x' \supset \bigwedge x' F^{*\prime}x'$ |
|---|---|---|
| Yes | Yes | Yes |
| No | No | Yes |

So for any arbitrary interpretation with a domain of just one member every sequence satisfies $F^{*\prime}x' \supset \bigwedge x' F^{*\prime}x'$. So the formula is 1-valid.

*Example 2*: Is $F^{*\prime}x' \supset \bigwedge x' F^{*\prime}x'$ 2-valid?

Let I be an arbitrary interpretation with a domain D of exactly two members, say $d_1$ and $d_2$. Let $s_1$ be an arbitrary sequence of members of D whose first term is $d_1$, $s_2$ an arbitrary sequence whose first term is $d_2$. Then in the case of $F^{*\prime}x'$ we get four possibilities:

| 1 | 2 |
|---|---|
| $s_1$ satisfies | $s_1$ does not satisfy |
| $s_2$ satisfies | $s_2$ satisfies |
| **3** | **4** |
| $s_1$ satisfies | $s_1$ does not satisfy |
| $s_2$ does not satisfy | $s_2$ does not satisfy |

For each of those four possibilities we work out whether or not both $s_1$ and $s_2$ satisfy the formula $F^{*\prime}x' \supset \bigwedge x' F^{*\prime}x'$, which we shall abbreviate to $W$.

Possibility 1: $s_1$ and $s_2$ both satisfy $F^{*\prime}x'$. Then both $s_1$ and $s_2$ satisfy $\bigwedge x' F^{*\prime}x'$, and so both satisfy $W$.

Possibility 2: $s_1$ does not, but $s_2$ does, satisfy $F^{*\prime}x'$. Then neither $s_1$ nor $s_2$ satisfies $\bigwedge x' F^{*\prime}x'$, and $s_1$ satisfies $W$ but $s_2$ does not.

This is enough to show that $W$ is not 2-valid. But for completeness we go on:

Possibility 3: $s_1$ satisfies $F^{*\prime}x'$ but $s_2$ does not. Then neither $s_1$ nor $s_2$ satisfies $\bigwedge x' F^{*\prime}x'$, and $s_2$ satisfies $W$ but $s_1$ does not.

Possibility 4: Neither $s_1$ nor $s_2$ satisfies $F^{*\prime}x'$. Then neither satisfies $\bigwedge x' F^{*\prime}x'$. But both satisfy $W$.

*Example 3*: Is $(F^{**\prime}a'a'' \supset (p' \supset \bigwedge x'' F^{**\prime}x'x''))$ 1-valid?

Abbreviate the formula to $Z$.

Let I be an arbitrary interpretation with a domain D of one member, d. Then there is only one denumerable sequence $s$ of members of D, viz. ⟨d, d, d, . . .⟩.

In the case of $F^{**\prime}x'x''$ there are just two possibilities: (1) $s$ satisfies it, (2) $s$ does not satisfy it.

In the case of $p'$ there are also just two possibilities: (1) $s$ satisfies it, (2) $s$ does not.

In the case of $F^{**\prime}a'a''$ there are the same two possibilities. But the possibilities for $F^{**\prime}a'a''$ and $F^{**\prime}x'x''$ are not independent: for $s$ satisfies $F^{**\prime}a'a''$ iff it satisfies $F^{**\prime}x'x''$ (the interpretation I must assign d to both $a'$ and a″: there is nothing else for it to assign).

This gives four relevantly different possibilities:

1. $s$ satisfies $F^{**\prime}x'x''$ (and so also $F^{**\prime}a'a''$) and also $p'$.
2. $s$ satisfies $F^{**\prime}x'x''$ but not $p'$.
3. $s$ does not satisfy $F^{**\prime}x'x''$ but does satisfy $p'$.
4. $s$ satisfies neither $F^{**\prime}x'x''$ nor $p'$.

Calculation will show that in all four cases $s$ satisfies $Z$. So $Z$ is 1-valid.

*Example 4*: Is the formula $Z$ (of Example 3 above) 2-valid?

Let I be an arbitrary interpretation with a domain of two

members, $d_1$ and $d_2$. Let $s_1, s_2, s_3, s_4$ be as in Illustration 2 above. Then for $F^{**\prime}x'x''$ we have 16 relevantly different possibilities, as in Illustration 2.

For each of those 16 possibilities there are four possible assignments that I might make to $a'$, $a''$, as in Illustration 4 above.

This gives $16 \times 4 = 64$ possibilities.

For each of those 64 possibilities the sequence under consideration may satisfy $p'$ or it may not.

This gives $64 \times 2 = 128$ relevantly different possibilities in all.

Calculation will show that not every sequence satisfies $Z$. For example, take the class of interpretations where

(i) $F^{**\prime}x'x''$ is satisfied by any sequence beginning $\langle d_1, d_1, \ldots \rangle$ but not by any sequence beginning $\langle d_1, d_2, \ldots \rangle$, and

(ii) $d_1$ is assigned to both $a'$ and $a''$, and

(iii) $p'$ is assigned the truth value T.

Let I be an interpretation in this class. Then, since by (i) the ordered pair $\langle d_1, d_1 \rangle$ has the property assigned by I to $F^{**\prime}$, by (ii) $F^{**\prime}a'a''$ is true for I. So any sequence satisfies $F^{**\prime}a'a''$ (for the interpretation I). Also, since $p'$ is true for I, any sequence satisfies $p'$ (for I). But by (i) any sequence beginning $\langle d_1, d_2, \ldots \rangle$ does not satisfy $F^{**\prime}x'x''$ and therefore does not satisfy $\bigwedge x'' F^{**\prime}x'x''$. So no sequence beginning $\langle d_1, d_2, \ldots \rangle$ satisfies $Z$ (for I). So $Z$ is not 2-valid.

Notice that we do not have to consider the particular nature of $d_1$ or $d_2$ or the particular property assigned by I to $F^{**\prime}$.

It is clear from these examples that there is an effective method for telling, for any formula of Q and any positive integer $k$, whether or not the formula is $k$-valid.

There are formulas that are $k$-valid for every positive integer $k$ but not logically valid: e.g. the formula (in abbreviated notation)

$$\sim(\bigwedge x \bigwedge y \bigwedge z(((Fxy \wedge Fyz) \supset Fxz) \wedge \sim Fxx) \wedge \bigwedge y \bigvee x Fxy)$$ [1]

So the existence of an effective method for determining the $k$-validity of formulas of Q does not imply the existence of an

[1] To see that it is not logically valid, take the natural numbers as domain and let $Fxy$ mean $x > y$.

effective method for determining their logical validity. In fact there is no effective method for determining their logical validity: cf. §57 below.

EXERCISE

1. (*a*) Is $\sim\Lambda x_1 \sim Fx_1 \supset \Lambda x_1 Fx_1$ [$\mathsf{V} x_1 Fx_1 \supset \Lambda x_1 Fx_1$] 1-valid?
   (*b*) Is it 2-valid?

ANSWER

1. (*a*) Yes. Let I be an arbitrary interpretation with a domain D of one member, d. Then there is just one denumerable sequence $s$ of members of D, viz. the sequence ⟨d, d, d, . . .⟩. Either $s$ satisfies $Fx$, or it does not. So we get the following table, where 'Yes' means '$s$ satisfies' and 'No' means '$s$ does not satisfy':

| $Fx_1$ | $\sim Fx_1$ | $\Lambda x_1 \sim Fx_1$ | $\sim\Lambda x_1 \sim Fx_1$ | $\Lambda x_1 Fx_1$ | $\sim\Lambda x_1 \sim Fx_1 \supset \Lambda x_1 Fx_1$ |
|---|---|---|---|---|---|
| Yes | No | No | Yes | Yes | Yes |
| No | Yes | Yes | No | No | Yes |

(*b*) No. Let $s_1$ and $s_2$ be as in Example 2 on p. 163. Abbreviate $\sim\Lambda x_1 \sim Fx_1 \supset \Lambda x_1 Fx_1$ to $Y$.

Possibility 1: $s_1$ and $s_2$ both satisfy $Fx_1$. Then they both satisfy $\Lambda x_1 Fx_1$ and therefore $Y$.

Possibility 2: $s_1$ does not, but $s_2$ does, satisfy $Fx_1$. Then neither $s_1$ nor $s_2$ satisfies $\Lambda x_1 Fx_1$. Since $s_2$ satisfies $Fx_1$, it does not satisfy $\sim Fx_1$; therefore neither $s_1$ nor $s_2$ satisfies $\Lambda x_1 \sim Fx_1$; therefore both satisfy $\sim\Lambda x_1 \sim Fx_1$. So both satisfy $\sim\Lambda x_1 \sim Fx_1$ and neither satisfies $\Lambda x_1 Fx_1$. So neither satisfies $Y$. So $Y$ is not 2-valid. We need not consider the remaining possibilities.

## 41 A deductive apparatus for Q: the formal system QS. Definitions of *proof in QS, theorem of QS, derivation in QS, syntactic consequence in QS,* [*proof-theoretically*] *consistent set of QS*

Treat this section as though it followed on directly from the end of §38 and as though you knew nothing about any semantics for Q.

*The system QS*

*Axioms*[1]

Let A, B, C be any wffs of Q (not necessarily distinct), v any individual variable, and t any term. If A is a wff, then At/v is a wff got from A by substituting t for all free occurrences of v in A. Then the following are axioms:

[QS1] $(A \supset (B \supset A))$
[QS2] $((A \supset (B \supset C)) \supset ((A \supset B) \supset (A \supset C))$
[QS3] $((\sim A \supset \sim B) \supset (B \supset A))$
[QS4] $(\Lambda vA \supset At/v)$ if t is free for v in A [i.e. if (1) if t is a variable, then t occurs free in At/v wherever v occurs free in A, and (2) if t is a term in which any variables occur, then wherever t is substituted for free occurrences of v in A all occurrences of variables in t remain free. If t is a closed term, t is free for v in A.]
[QS5] $(A \supset \Lambda vA)$ if v does not occur free in A
[QS6] $(\Lambda v(A \supset B) \supset (\Lambda vA \supset \Lambda vB))$

In addition:

[QS7] If A is an axiom, then $\Lambda vA$ is also an axiom.

The reason for the restriction on QS4 is that we want all our axioms to be logically valid and without the restriction we should have the following formula, for example, as an axiom:

$$\Lambda x_1 \sim \Lambda x_2 Fx_1x_2 \supset \sim \Lambda x_2 Fx_2x_2$$

That formula is not logically valid. (Let '*F*' mean ' = ' and let D have at least two members. Then the antecedent is true and the consequent false.)

The reason for the restriction on QS5 is that without it we should have the following formula, for example, as an axiom:

$$Fx_1 \supset \Lambda x_1 Fx_1$$

That formula is not logically valid. (Take D as the set of natural numbers and '*F*' as 'is even'. Then the sequence ⟨2, 2, 2, 2, . . .⟩, for example, does not satisfy the formula.)

[1] This set of axioms is chosen because with it we get

$$\Gamma \vdash_{QS} A \quad \text{iff} \quad \Gamma \vDash_Q A \qquad [46.3]$$

Some other sets of axioms for predicate logic do not have this property.

*Rule of inference*

If A and B are wffs then B is an immediate consequence of A and $(A \supset B)$.

[Modus Ponens for $\supset$]

*Definition.* A *proof in QS* is a finite string of formulas of Q, each one of which is either an axiom of QS or an immediate consequence by the rule of inference of QS of two formulas preceding it in the string.

*Definition.* A formula A of Q is a *theorem of QS* [$\vdash_{QS} A$] iff there is some proof in QS whose last formula is A.

*Definition.* A string of formulas is a *derivation in QS* of a wff A from a set Γ of wffs of Q iff (1) it is a finite (but not empty) string of formulas of Q, (2) the last formula in the string is A, and (3) each formula in the string is either (i) an axiom of QS or (ii) an immediate consequence by the rule of inference of QS of two formulas preceding it in the string or (iii) a member of the set Γ.

*Definition.* A formula A is a *syntactic consequence in QS* of a set Γ of formulas of Q [$\Gamma \vdash_{QS} A$] iff there is a derivation in QS of A from the set Γ.

*Definition.* A set Γ of formulas of Q is a *proof-theoretically consistent set of QS* iff for no formula A of Q is it the case that both $\Gamma \vdash_{QS} A$ and $\Gamma \vdash_{QS} \sim A$.

We shall abbreviate 'proof-theoretically consistent set' to 'consistent set', because we shall not be using the phrase 'model-theoretically consistent set [of Q]'. [Sometimes we shall speak of a set as *having a model*, sometimes of a set as *being simultaneously satisfiable*, these being distinct notions.]

A set Γ of formulas of Q is an *inconsistent set of QS* iff for some formula A of Q both $\Gamma \vdash_{QS} A$ and $\Gamma \vdash_{QS} \sim A$.

## 42 Proof of the consistency of QS

42.1 *QS is consistent*

*Proof.* Let A be any wff of Q. We define its associated propositional formula, $A^{prop}$, as follows: Delete all quantifiers in

A; delete all terms in A; replace each predicate symbol by the propositional symbol *p*′. $A^{prop}$ is a wff of the language P.

If A is an axiom of QS by one of the axiom-schemata QS1, QS2, QS3, then $A^{prop}$ will be an axiom of PS by one of the axiom-schemata [of PS] PS1, PS2, PS3. If A is an axiom of QS by QS4, QS5 or QS6, then $A^{prop}$ will be a theorem of PS of the form $(B \supset B)$. If A is an axiom of QS by axiom-schema QS7, then $A^{prop}$ will be either an axiom of PS or a theorem of PS of the form $(B \supset B)$.

If B is an immediate consequence by the rule of inference of QS of A and $(A \supset B)$, then $B^{prop}$ will be an immediate consequence by the rule of inference of PS of $A^{prop}$ and $(A \supset B)^{prop}$; for $(A \supset B)^{prop}$ is $(A^{prop} \supset B^{prop})$.

Now suppose that QS is inconsistent. Then for some wff A of Q there is a proof in QS of A and also a proof in QS of $\sim A$. For each wff in these proofs substitute its associated propositional formula. The result will be two strings of wffs of P. If either of these strings is not a proof in PS, this will be because at some place or other a wff of the form $(B \supset B)$ occurs in the string without being an immediate consequence in PS of wffs preceding it in the string. Put in extra steps in the string(s) to produce proofs in PS of any such wffs [we know this can be done, since $(B \supset B)$ is a theorem-schema of PS] and the result will be proofs *in PS* of $A^{prop}$ and $\sim A^{prop}$, where $\sim A^{prop}$ is the negation of $A^{prop}$.

So we have: If QS is inconsistent then so is PS.

But PS is consistent. So QS is.

Variant of the proof: Any theorem of QS will have a tautology of P as its associated propositional formula [any axiom of QS has a tautology of P as its a.p.f., and Modus Ponens in QS preserves tautologousness of associated propositional formulas]. Now *F**′*x*′ is a wff of Q that does not have a tautology of P as its a.p.f. Therefore it is not a theorem of QS. Therefore QS is consistent.

Another proof of the consistency of QS can be got by using 43.5 below: If $\vdash_{QS} A$, then $\vDash_{Q} A$. [However, there is a sense of 'finitary' in which the proofs of consistency given above are finitary, while the proof of 43.5 is not: see further Kleene (1952, pp. 174–5).]

## 43 Some metatheorems about QS

We develop here only as much of the metatheory of QS as we need for later proofs.

43.1 *The Deduction Theorem holds for QS*

*Proof.* By our analysis of the proof of the Deduction Theorem for PS at the end of §26, the DT holds for any system having the three properties:

1. Any wff of the form $A \supset (B \supset A)$ is a theorem.
2. Any wff of the form $(A \supset (B \supset C)) \supset ((A \supset B) \supset (A \supset C))$ is a theorem.
3. Modus Ponens for $\supset$ is the sole rule of inference.

QS has these three properties (1 by QS1, 2 by QS2).

43.2 (*The converse of the Deduction Theorem*: *used in Part 4*)
*If $\Gamma \vdash_{QS} A \supset B$, then $\Gamma, A \vdash_{QS} B$*

*Proof.* Suppose $\Gamma \vdash_{QS} A \supset B$. Then $\Gamma, A \vdash_{QS} A \supset B$. But $\Gamma, A \vdash_{QS} A$. So by Modus Ponens $\Gamma, A \vdash_{QS} B$.

43.3 *If $\Delta$ is a set of closed wffs, then if $\Delta \vdash_{QS} A$ then $\Delta \vdash_{QS} \Lambda vA$*

*Proof.* We show how, given a derivation of A from $\Delta$, to find a derivation of $\Lambda$vA from $\Delta$. The proof is by induction on the length, *n*, of the derivation of A from $\Delta$.

*Basis*: $n = 1$

Then A is either an axiom or in $\Delta$. If A is an axiom then $\Lambda$vA is also an axiom, by QS7, and so the required derivation consists simply of the single formula $\Lambda$vA. If A is in $\Delta$, then the following string is a derivation of $\Lambda$vA from $\Delta$:

A
$A \supset \Lambda vA$ [QS5. Since A is closed, v does not occur free in A.]
$\Lambda vA$

*Induction Step*

Assume the theorem holds for all derivations of A from $\Delta$ of length less than *k*. To prove it holds for all derivations of length *k*. 3 cases:

1. A is an axiom
2. A is in $\Delta$
3. A is an immediate consequence by MP of two preceding formulas

*Cases 1 and 2* are as in the Basis.

*Case 3.* A is an immediate consequence by MP of two formulas preceding it in the derivation $A_1, \ldots, A_k$ of A from $\Delta$ ($A_k$ is A). Let these two formulas be $A_i$, $A_j$, where $i<k$ and $j<k$ and $A_j$ is $A_i \supset A$. Then by the induction hypothesis there are derivations of $\Lambda v A_i$ and $\Lambda v(A_i \supset A)$ from $\Delta$. By QS6 $\vdash_{QS} \Lambda v(A_i \supset A) \supset (\Lambda v A_i \supset \Lambda v A)$. So by two applications of Modus Ponens we get $\Delta \vdash_{QS} \Lambda v A$.

Reminder: If A is a tautology of P whose only propositional symbols are $P_1, \ldots, P_n$, then the result of substituting wffs $Q_1, \ldots, Q_n$ of Q for $P_1, \ldots, P_n$ respectively in A is an *instance of a tautological schema of Q*.

43.4 *If A is an instance of a tautological schema of Q, then* $\vdash_{QS} A$

*Proof.* Suppose that A is an instance of a tautological schema of Q. Let B be a tautology of P of which A is a substitution instance. Then B has a proof in PS (by the semantic completeness theorem for PS). Since A is the result of substituting wffs of Q for the propositional symbols of B, we make the same substitutions throughout the proof of B. If any propositional symbol in the proof of B does not occur in B itself, replace it by an arbitrary wff of Q throughout the proof. Under this substitution axioms of PS by PS1–3 become axioms of QS by QS1–3 and uses of Modus Ponens in PS become uses of Modus Ponens in QS. So we get a proof of A in QS.

43.5 *If* $\vdash_{QS} A$ *then* $\models_Q A$ [*i.e. every theorem of QS is logically valid*]

*Proof.* By 40.10 every axiom by any of QS1–3 is logically valid. By 40.16 every axiom by QS4 is logically valid. By 40.13 every axiom by QS5 is logically valid. By 40.11 every axiom by QS6 is logically valid. So every axiom by any of QS1–6 is logically valid. So by 40.6 every axiom by QS7 applied to any axiom by QS1–6 is logically valid. So again by 40.6 any axiom by QS7 is logically valid. So every axiom of QS is logically valid. By 40.4 Modus Ponens preserves logical validity. So every theorem of QS is logically valid.

This result yields another proof of the consistency of QS [$p'$ is a wff of Q that is not logically valid, so not a theorem. So QS is consistent].

43.6 *If* $\Gamma \vdash_{QS} A$, *then there is a finite subset* $\Delta$ *of* $\Gamma$ *such that* $\Delta \vdash_{QS} A$

This follows from our requirement that a derivation in QS must be a *finite* string of formulas.

43.7 *If* $\Gamma \vdash_{QS} A$, *then* $\Gamma \vDash_{Q} A$

*Proof.* Suppose that $\Gamma \vdash_{QS} A$. Then by 43.6 there is a finite subset $\Delta$ of $\Gamma$ such that $\Delta \vdash_{QS} A$. (i) If $\Delta$ is empty, then $\vdash_{QS} A$, and so by 43.5 $\vDash_{Q} A$, and so $\Gamma \vDash_{Q} A$. (ii) If $\Delta$ is not empty, let $A_1, \ldots, A_n$ be the members of $\Delta$. Then $A_1, \ldots, A_n \vdash_{QS} A$. So by the DT, applied as many times as may be necessary, $\vdash_{QS} A_1 \supset (A_2 \supset (\ldots\ldots (A_n \supset A) \ldots))$. So by 43.5 $\vDash_{Q} A_1 \supset (A_2 \supset (\ldots\ldots (A_n \supset A) \ldots))$. So there is no sequence that satisfies $\{A_1, \ldots, A_n\}$ and that does not also satisfy A, i.e. there is no sequence that satisfies every member of $\Delta$ that does not also satisfy A. Since $\Delta$ is a subset of $\Gamma$, it follows that there is no sequence that satisfies every member of $\Gamma$ and that does not also satisfy A: i.e. $\Gamma \vDash_{Q} A$.

43.8 *If* $A \vdash_{QS} B$, *then* $\vDash_{Q} A \supset B$

*Proof.* Suppose $A \vdash_{QS} B$. Then by the DT $\vdash_{QS} A \supset B$. Then by 43.5 $\vDash_{Q} A \supset B$.

43.9 *If* $\Gamma \cup \{\sim A\}$ *is an inconsistent set of QS, then* $\Gamma \vdash_{QS} A$

*Proof.* Suppose $\Gamma \cup \{\sim A\}$ is an inconsistent set of QS. Then $\Gamma, \sim A \vdash_{QS} B$ and $\Gamma, \sim A \vdash_{QS} \sim B$ for some formula B. So by the DT $\Gamma \vdash_{QS} \sim A \supset B$ and $\Gamma \vdash_{QS} \sim A \supset \sim B$. But $\vdash_{QS} (\sim A \supset B) \supset ((\sim A \supset \sim B) \supset A)$ (instance of a tautological schema). So by MP twice $\Gamma \vdash_{QS} A$.

In the next section (§44) we define the notion of a *first order theory*, and in the long section after that (§45) we prove various metatheorems about arbitrary first order theories. It is through first order theories that modern logic finds most of its applications. Set theory, for example, can be presented as an interpreted first order theory, and practically all (? all) pure mathematics can be expressed in the language of set theory. But we are interested in first order theories for a different reason. Our main purpose is to prove Theorem 45.15, *Any consistent first order theory has a denumerable model*, from which the semantic completeness of QS follows by a short and simple argument. Theorem 45.15 is in fact the key theorem in this part of the book. Its proof is long and sometimes painful, but it has such interesting consequences that the pains are well worth while.

## 44 First order theories

A *first order theory* is a formal system that satisfies the following conditions:

1. Its language is Q or $Q^+$ [$Q^+$ is Q with the addition of denumerably many new individual constants, there being given an effective enumeration of the added constants].
2. Its *logical axioms* are specified by the schemata QS1–7 applied to the wffs of its language, with QS7 re-worded to 'If A is a *logical* axiom, then ΛvA is also a *logical* axiom'. So every first order theory has for its logical axioms all the axioms of QS (and more, if its language is $Q^+$).
3. It may have in addition countably many *proper axioms*, which are to be *closed* wffs of its language.
4. Modus Ponens for ⊃ is its sole rule of inference.

It follows from this definition that QS itself is a first order theory.

Let K be an arbitrary first order theory. Then we shall call the axiom-schemata of K corresponding to the schemata QS1–7 K1–7.

Let K be an arbitrary first order theory. Then the definitions of *proof in K*, *theorem of K*, *derivation in K*, *syntactic consequence in K*, *consistent set of K* are exactly like those for QS except that throughout those definitions both 'QS' and 'Q' are replaced by 'K'. (A formula of K is a formula of the language of K, and this will be either Q or $Q^+$, depending on what K is.)

A *model* of a first order theory K is a model of the set of theorems of K.

## 45 Some metatheorems about arbitrary first order theories. Negation-completeness. Closed first order theories. The Löwenheim–Skolem Theorem. The Compactness Theorem

Throughout this section let K be an arbitrary first order theory. N.B. All proper axioms of K are *closed* wffs.

45.1 *The Deduction Theorem holds for any arbitrary first order theory*

*Proof.* Any arbitrary first order theory has the three properties mentioned at the end of §26. Cf. 43.1.

45.2 *The converse of the Deduction Theorem holds for any arbitrary first order theory*

*Proof.* Suppose $\Gamma \vdash_K A \supset B$. Then $\Gamma$, $A \vdash_K A \supset B$. But $\Gamma$, $A \vdash_K A$. So by Modus Ponens $\Gamma$, $A \vdash_K B$.

45.3 *Let K be an arbitrary first order theory. Any instance, in the language of K, of a tautological schema is a theorem of K*

*Proof.* As for 43.4, replacing 'QS' and 'Q' by 'K' throughout.

45.4 *If* $\vdash_K A$, *then* $\vdash_K \Lambda v A$

*Proof.* By induction on the length, *n*, of a proof in K of A.

*Basis*: $n = 1$

Then A is either a logical or a proper axiom of K. If it is a logical axiom of K, then $\Lambda$vA is also a logical axiom of K, by K7. If it is a proper axiom of K, then $\Lambda$vA is a theorem of K, by the steps

$A$
$A \supset \Lambda v A$ [K5: A is closed]
$\Lambda v A$

*Induction Step*

Assume the theorem holds for all proofs of length less than *k*. To prove it holds for all proofs of length *k*.
2 cases:

1. A is an axiom

2. A is an immediate consequence by MP of two preceding wffs

*Case 1.* As in the Basis.

*Case 2.* Let $A_1, \ldots, A_k$ be a proof in K of A (so $A_k$ is A). A is an immediate consequence by MP of $A_i$ and $A_j$ where $i<k$ and $j<k$ and $A_j$ is $A_i \supset A$. Then by the induction hypothesis $\vdash_K \Lambda vA_i$ and $\vdash_K \Lambda v(A_i \supset A)$. By K6 $\vdash_K \Lambda v(A_i \supset A) \supset (\Lambda vA_i \supset \Lambda vA)$. So by MP twice $\vdash_K \Lambda vA$.

45.5 $\vdash_K A$ *iff* $\vdash_K A^c$

*Proof.* Suppose $\vdash_K$ A. Then by 45.4 $\vdash_K \Lambda vA$ for arbitrary v. So, using 45.4 as often as may be necessary, $\vdash_K A^c$. Conversely, suppose $\vdash_K A^c$. Then, using K4 in the form $\Lambda vA \supset A$ as often as may be necessary, we get $\vdash_K$ A.

*Notation*

Let K be an arbitrary first order theory. Let A be a wff of K. Then *K* + {A} is to be the system that results when A is added to K as an extra axiom ['added' and 'extra' here are not intended to exclude the case where A is already an axiom of K].

45.6 (*a*) *If A is a closed wff of K that is not a theorem of K, then K* + {$\sim$*A*} *is a consistent first order theory*

*Proof.* Abbreviate K + {$\sim$A} to K′. Since K is a first order theory and $\sim$A is closed, K′ is a first order theory; and clearly $\vdash_{K'}$ B iff $\sim A \vdash_K B$, for any wff B. Now suppose K′ is inconsistent. Then $\vdash_{K'}$ C and $\vdash_{K'} \sim C$ for some wff C. Therefore $\sim A \vdash_K C$ and $\sim A \vdash_K \sim C$. Therefore by the Deduction Theorem $\vdash_K \sim A \supset C$ and $\vdash_K \sim A \supset \sim C$. But $\vdash_K (\sim A \supset C) \supset ((\sim A \supset \sim C) \supset A)$. Therefore by MP twice $\vdash_K$ A. But this contradicts our hypothesis that A was not a theorem of K.

45.6 (*b*) As for (*a*), but interchanging A and $\sim$A. The proof uses $\vdash_K (A \supset C) \supset ((A \supset \sim C) \supset \sim A)$.

45.7 (*a*) (Alternative version of 45.6 (*a*)) *If A is a closed wff of K that is not a theorem of K, then* {$\sim$*A*} *is a consistent set of K*

*Proof.* Suppose {$\sim$A} is an inconsistent set of K. Then $\sim A \vdash_K C$ and $\sim A \vdash_K \sim C$ for some C. The rest of the proof is as for 45.6 (*a*).

45.7 (*b*) As for 45.7 (*a*), but interchanging A and ~A.

45.8 *If a first order theory has a model, then it is consistent*

*Proof.* Let K be an arbitrary first order theory. Suppose K has a model M. And suppose K is inconsistent. Then $\vdash_K$ A and $\vdash_K$ ~A for some formula A. Then A is true for M and ~A is true for M. But this is impossible. So if K has a model K is consistent.

45.9 *For any first order theory K there are effective enumerations of*

1. *The wffs of K*
2. *The closed wffs of K*
3. *The wffs of K in which exactly one variable occurs free*
4. *The closed terms of K*

*Proof*

A. Suppose K has language Q. Assign numerals to the symbols of Q as follows:

| | |
|---|---|
| *p* | 10 |
| ′ | 100 |
| *x* | 1000 |
| *a* | 10000 |
| *f* | 100000 |
| *F* | 1000000 |
| * | 10000000 |
| ~ | 100000000 |
| ⊃ | 1000000000 |
| ⋀ | 10000000000 |
| ( | 100000000000 |
| ) | 1000000000000 |

1. The rest of the proof is exactly as for the Enumeration Theorem for the wffs of the language P. Let the resulting enumeration of the wffs of Q be E.

2. Delete from E all wffs in which any variable occurs free.

3. Delete from E all wffs other than wffs in which exactly one variable occurs free.

4. As for 1, but listing closed terms instead of wffs.

B. Suppose K has language Q+. Assign to the first new constant in the given effective enumeration of the new constants the numeral 10000000000000, i.e. '1' followed by 13 '0's. To the second constant in the enumeration assign the numeral consisting of '1' followed by 14 '0's, to the third the numeral consisting of '1' followed by 15 '0's. And so on. The rest of the proof is as for A.

*Definition.* A system S′ is an *extension* of a system S iff every theorem of S is a theorem of S′. N.B. If S′ is an extension of S, then every model of S′ is a model of S.

*Definition.* A system S is *negation-complete* iff for every *sentence* [closed wff] A of S either A or the negation of A is a theorem of S.

45.10 (*Lindenbaum's Lemma for first order theories*) *If K is a consistent first order theory, then there is a first order theory K′ that is a consistent negation-complete extension of K with the same formulas as K*

*Proof.* Let K be a consistent first order theory. By 45.9 there is an enumeration of the closed wffs of K. Let $\langle C_1, C_2, C_3, \ldots \rangle$ be such an enumeration. We define an infinite sequence, $\langle K_0, K_1, K_2, \ldots \rangle$, of first order theories as follows:

Let $K_0$ be K. If $\sim C_{n+1}$ is not a theorem of $K_n$, then $K_{n+1}$ is to be $K_n + \{C_{n+1}\}$ [i.e. $K_n$ with $C_{n+1}$ added as an axiom]. If $\sim C_{n+1}$ is a theorem of $K_n$, then $K_{n+1}$ is to be $K_n$. [I.e. if $C_{n+1}$ can consistently be added to $K_n$, add it to $K_n$ to make $K_{n+1}$. If it cannot, let $K_{n+1}$ be $K_n$.]

Let $K_\infty$ be the system that results from taking as axioms all the axioms of all the $K_i$'s. Clearly $K_\infty$ is an extension of K with the same formulas as K and also a first order theory. We show that it is consistent and negation-complete.

*$K_\infty$ is consistent*

1. Each of the $K_i$'s is consistent. For $K_0$ is a consistent first order theory *ex hypothesi*, and $K_1$ is either $K_0$ or, by 45.6 (*b*), a consistent first order theory. So $K_1$ is a consistent first order theory. Similarly $K_2$ is a consistent first order theory. And so on.

2. $K_\infty$ is consistent. For suppose not. Then there is a proof in $K_\infty$ of A and a proof in $K_\infty$ of $\sim$A for some formula A. Each of these proofs consists of only finitely many formulas, and so uses only finitely many of the closed wffs that were added as axioms to K to make $K_\infty$. Each of these closed wffs has a

number in our enumeration. Let the highest-numbered closed wff used as an axiom in either proof be $C_n$. Then there is a proof in $K_n$ of A and a proof in $K_n$ of $\sim$A; i.e. $K_n$ is inconsistent. This contradicts 1 above. So $K_\infty$ is consistent.

*$K_\infty$ is negation-complete*

Let $C_m$ be the *m*th closed wff in our enumeration. Then either $\sim C_m$ is a theorem of $K_{m-1}$ (and therefore of $K_\infty$) or $C_m$ is a theorem of $K_m$ (and therefore of $K_\infty$). This holds for each positive integer *m*, and so for each closed wff of K.

*Remark.* Though the enumeration of the closed wffs of K is effective, we shall see later (Part 4) that there is no effective method of telling, for arbitrary closed wffs of Q (or $Q^+$), whether or not they are derivable in QS from arbitrary sets of wffs of Q (or $Q^+$). So our definition of $K_\infty$ does not provide us with an effective method for telling whether or not a formula is an axiom of $K_\infty$. The set of axioms of $K_\infty$ might not be decidable. This is why we did not require that a formal system should, by definition, have a decidable set of proofs (p. 16 above).

45.11 *If K is a consistent first order theory, then the system that results from adding a denumerable set of new individual constants to K, with an effective enumeration of those constants, is a consistent first order theory that is an extension of K*

*Proof.* Let K′ be the system with added constants. K has at most an effectively enumerable set of individual constants that are not in Q. K′ has an effectively enumerable set of individual constants that are not in Q. The union of two effectively enumerable sets is still an effectively enumerable set (take members from each set alternately). So any individual constants in K′ but not in Q are effectively enumerable. K′ has its logical axioms defined by the schemata K′ 1–7. K′ has exactly the same proper axioms as K (adding new constants adds new logical axioms but no new proper axioms). So since K is a first order theory, K′ is. Obviously K′ is an extension of K. It remains to prove that K′ is consistent.

Suppose that K′ is inconsistent. Then for some formula A there is a proof in K′ of A and also a proof in K′ of $\sim$A. Each of these proofs consists of only finitely many formulas and so uses only finitely many of the new constants (and also only finitely many variables). Let $b_1, \ldots, b_n$ be all the distinct new constants that occur in these proofs. Let $u_1, \ldots, u_n$ be distinct variables that do not occur in either of the proofs. We shall show that the strings of formulas that result from replacing $b_1, \ldots, b_n$ by $u_1, \ldots, u_n$ respectively throughout the proofs are proofs in K of $Au_i/b_i$ and of $\sim Au_i/b_i$, where $Au_i/b_i$ is the result of replacing each new constant that occurs in A by its corresponding variable from the list $u_1, \ldots, u_n$. So if K′ is inconsistent, K is. But K is consistent *ex hypothesi*. So K′ is consistent.

Let $A_1, \ldots, A_k$ be a proof in K′ of A (so $A_k$ is A). Let $A_1', \ldots, A_k'$ be the result of replacing the new constants in $A_1, \ldots, A_k$ by variables from the list $u_1, \ldots, u_n$, as explained in the last paragraph. We want to show that $A_1', \ldots, A_k'$ is a proof in K of A′, i.e. of $Au_i/b_i$. Now for each $A_i$ there are three possibilities:

*Case 1.* $A_i$ is an axiom of K′ by one or other of the schemata K′ 1–7. Anything that is an axiom by any of K′ 1–7 and has one or more of the new constants in it will be an axiom of K if the distinct new constants are replaced by distinct variables that do not occur elsewhere in the axiom. So $A_i'$ will be an axiom of K.

*Case 2.* $A_i$ is a proper axiom of K′. K′ has no proper axioms other than those of K. So $A_i$ is a proper axiom of K, and therefore has no new constants in it. So $A_i'$ is $A_i$, a proper axiom of K.

*Case 3.* $A_i$ is an immediate consequence by Modus Ponens of two formulas $A_j$ and $A_j \supset A_i$ preceding it in the proof. Then $A_i'$ will be an immediate consequence by Modus Ponens of $A_j'$ and $(A_j \supset A_i)'$, for $(A_j \supset A_i)'$ is simply $(A_j' \supset A_i')$.

So if $A_1, \ldots, A_k$ is a proof in K′ of A, $A_1', \ldots, A_k'$ is a proof in K of $Au_i/b_i$.

By an exactly similar argument, if there is a proof in K′ of $\sim$A, then there is a proof in K of $\sim Au_i/b_i$.

45.12 *Let $\Lambda vA$ be a closed wff of K, and let c be a constant of K that does not occur in any proper axiom of K. Let K′ be the system that results from adding $Ac/v \supset \Lambda vA$ to K as a proper axiom. If K is a consistent first order theory, then K′ is a consistent first order theory that is an extension of K*

*Proof.* It is obvious that K′ is a first order theory and an extension of K. Now suppose that K′ is inconsistent. Then for some formula A there is a proof in K′ of A and a proof in K′ of ~A. Let u be a variable that does not occur in either of these proofs. Replace c by u throughout both proofs. We shall show that the result will be proofs in K of Au/c and ~Au/c. So if K′ is inconsistent, so is K. So if K is consistent so is K′.

Let $A_1, \ldots, A_k$ be a proof in K′ of A. Let $A_1', \ldots, A_k'$ be the result of replacing c by u throughout the proof. For each $A_i$ there are four possibilities:

*Case 1.* $A_i$ is a logical axiom of K′ (in which, by the definition of u, u does not occur). Then $A_i'$ is a logical axiom of K.

*Case 2.* $A_i$ is a proper axiom of K. Then, since *ex hypothesi* c does not occur in any proper axiom of K, $A_i'$ is $A_i$.

*Case 3.* $A_i$ is $Ac/v \supset \Lambda vA$. Now the formula that results from replacing all occurrences of c in Ac/v by u is the same as the formula that results from replacing all occurrences of c and all free occurrences of v in A by u. Let this formula be B. Then $A_i'$ is $B \supset \Lambda vB$, where v does not occur free in B; and this is a logical axiom of K by K5.

*Case 4.* $A_i$ is an immediate consequence by MP of two formulas $A_j$ and $A_j \supset A_i$ preceding it in the proof. Then $A_i'$ will be an immediate consequence by MP of $A_j'$ and $(A_j \supset A_i)'$, since $(A_j \supset A_i)'$ is $(A_j' \supset A_i')$.

So if $A_1, \ldots, A_k$ is a proof in K′ of A, then $A_1', \ldots, A_k'$ is a proof in K of Au/c. By an exactly similar argument, if there is a proof in K′ of ~A, then there is a proof in K of ~Au/c.

*Definition.* A formal system S that has at least one closed term is *closed* iff for each formula A with just one free variable, if the result of substituting a closed term for the free occurrences of the variable is a theorem of S for *every* closed term of S, then $\Lambda vA$ (where v is the variable in question) is also a theorem of S.

(Informally, this means that in any model of the system, if something is true of every member of the domain that has a name in the system, then it is true of *every* member of the domain: i.e. in every model of the system every member of the domain has a name in the system.)

45.13 *If K is a consistent first order theory, then there is a first order theory K′ that is a consistent, closed, and negation-complete extension of K*

*Proof.* Let K be an arbitrary consistent first order theory. Let $K_0$ be the system that results from K by adding a denumerable set of new individual constants, with an effective enumeration of them. By 45.11 $K_0$ is a consistent first order theory. By 45.9 (3) there is an enumeration of the wffs of $K_0$ in which exactly one variable occurs free. Let $\langle A_1, A_2, A_3, \ldots\rangle$ be such an enumeration. Let $\langle b_1, b_2, b_3, \ldots\rangle$ be an enumeration of the individual constants that were added to K to convert it into $K_0$. Let $S_1$ be the wff

$$A_1 b_{j_1}/v_1 \supset \Lambda v_1 A_1$$

where $v_1$ is the variable that occurs free in $A_1$, and $b_{j_1}$ is the first constant in our enumeration $\langle b_1, b_2, b_3, \ldots\rangle$ that does not occur in $A_1$. For each positive integer $n$, $n > 1$, let $S_n$ be the wff

$$A_n b_{j_n}/v_n \supset \Lambda v_n A_n$$

where $b_{j_n}$ is the first constant in our enumeration that does not occur in any of $S_1, \ldots, S_{n-1}$ or in $A_n$.

Let $K_1$ be the system that results from adding $S_1$ as a proper axiom to $K_0$; $K_{n+1}$ the result of adding $S_{n+1}$ to $K_n$. Let $K_\infty$ be the system that results from adding all the denumerably many $S_i$'s to $K_0$ as proper axioms. Then $K_\infty$ is obviously an extension of K, and a first order theory (in particular, all its proper axioms are closed wffs). We prove next that $K_\infty$ is consistent and closed.

*$K_\infty$ is consistent*

1. Each $K_i$ is consistent. For $K_0$ is consistent, and by 45.12 if $K_n$ is consistent, then so is $K_{n+1}$.

2. $K_\infty$ is consistent. For suppose not. Then for some formula A there would be proofs in $K_\infty$ of A and $\sim$A. These proofs would consist of only finitely many formulas, and so would use only finitely many of the $S_i$'s. Suppose the highest-numbered $S_i$ used in either proof is $S_n$. Then there would be proofs in $K_n$ of A and of $\sim$A. So one of the $K_i$'s would be inconsistent, which contradicts 1 above. So $K_\infty$ is consistent.

*$K_\infty$ is closed*

Suppose $A_k c/v_k$ is a theorem of $K_\infty$ for every closed term c of $K_\infty$ ($v_k$ is the variable that occurs free in $A_k$). Then it is a theorem for the constant $b_{j_k}$ that occurs in the axiom $S_k$ of $K_\infty$, viz. the axiom

$$A_k b_{j_k}/v_k \supset \Lambda v_k A_k$$

and therefore by Modus Ponens we get

$$\vdash_{K_\infty} \Lambda v_k A_k$$

This holds for each positive integer $k$. So for each formula A with just one free variable, if the result of substituting a closed term for the free occurrence(s) of the variable is a theorem of $K_\infty$ for every closed term of $K_\infty$, then so is $\Lambda$vA, where v is the variable in question. So $K_\infty$ is closed.

By 45.10 there is a first order theory K′ that is a consistent negation-complete extension of $K_\infty$ with the same formulas as $K_\infty$. K′ will be closed, sincc $K_\infty$ is and K′ has the same formulas as $K_\infty$. So K′ is a first order theory that is a consistent, closed, and negation-complete extension of K. (K was an arbitrary consistent first order theory.)

*Definitions.* A *model* of a formal system is said to be *finite* / *denumerable* / *countable* / *uncountable* according as the domain of the model is finite / denumerable / countable / uncountable.

45.14 *Any consistent, closed, negation-complete first order theory has a denumerable model*

*Proof.* Let T be any consistent, closed, negation-complete first order theory. Let I be the following interpretation of T:

1. The domain D of I is the set of *closed terms* of T. This is denumerable, by 45.9.

2. I assigns to each constant c of T c itself (as its own interpretation, so to speak).

3. To each $n$-place function symbol $f_i$ I assigns the function $f_i$, with arguments and values in D, defined by the rule:

The value of $f_i(x_1, \ldots, x_n)$ for the arguments $x_1 = t_1, \ldots, x_n = t_n$, where $t_1, \ldots, t_n$ are closed terms of T, is the closed term $f_i t_1 \ldots t_n$.

4. To each $n$-place predicate symbol $F_i$ I assigns the set of all ordered $n$-tuples $\langle t_1, \ldots, t_n \rangle$ of closed terms of T such that $\vdash_T F_i t_1 \ldots t_n$.

5. To each propositional symbol $p_i$ I assigns the truth value truth iff $\vdash_T p_i$.

We shall prove that a formula A of T is true for I iff $\vdash_T$ A, and hence that I is a model of T. We need only consider closed wffs, since a formula A is true for I iff $A^c$ is (40.7) and $\vdash_T$ A iff $\vdash_T A^c$ (45.5).

Lemma to be proved: A closed wff A is true for I iff $\vdash_T$ A. The proof is by induction on the number, *n*, of quantifiers and connectives in A.

*Basis*: $n = 0$

Then A is either a propositional symbol or of the form $Ft_1 \ldots t_m$, where $t_1, \ldots, t_m$ are closed terms (A is a closed wff). In either case A is true for I iff $\vdash_T$ A, by the assignments in clauses 5 and 4 respectively of our specification of I.

*Induction Step*

Assume the lemma holds for all closed wffs with fewer than *k* connectives and quantifiers. To prove that it holds for all closed wffs with exactly *k* connectives and quantifiers. 3 possibilities:

1. A is $\sim$B
2. A is $B \supset C$
3. A is $\Lambda v_j B$

*Case 1.* A is $\sim$B. Since A is closed, B is.

1. Suppose A is not true for I. Then B is true for I. So by the induction hypothesis $\vdash_T$ B. So by the consistency of T $\sim$B is not a theorem of T: i.e. A is not a theorem of T. So if $\vdash_T$ A, then A is true for I.

2. Suppose A is true for I. Then B is false for I. So by the induction hypothesis B is not a theorem of T. So by the negation-completeness of T $\vdash_T \sim$B: i.e. $\vdash_T$ A. So if A is true for I, $\vdash_T$ A.

*Case 2.* A is $B \supset C$. Since A is closed, B and C are too.

1. Suppose A is not true for I. Then B is true for I and C is false for I. So by the induction hypothesis $\vdash_T$ B and C is not a theorem of T. So by the negation-completeness of T $\vdash_T \sim$C. So by the tautology $B \supset (\sim C \supset \sim(B \supset C))$, which is a theorem of T, we get $\vdash_T \sim(B \supset C)$, i.e. $\vdash_T \sim$A. So by the consistency of T A is not a theorem of T. So if A is a theorem of T, then A is true for I.

2. Suppose A is not a theorem of T. Then by the negation-completeness of T $\vdash_T \sim A$, i.e. $\vdash_T \sim(B \supset C)$. By $\vdash_T \sim(B \supset C) \supset B$ and $\vdash_T \sim(B \supset C) \supset \sim C$ we get $\vdash_T B$ and $\vdash_T \sim C$. So by the induction hypothesis B is true for I. By the consistency of T C is not a theorem of T; so by the induction hypothesis C is not true for I. Since C is closed C is false for I. So $B \supset C$, i.e. A, is not true for I. So if A is true for I, $\vdash_T A$.

*Case 3.* A is $\Lambda v_j B$. Two possibilities: (I) B is closed. (II) B is open.

I. B is closed.

1. Suppose $\vdash_T A$. Then $\vdash_T B$ (by K4 and Modus Ponens). So by the induction hypothesis B is true for I. So by 40.6 $\Lambda v_j B$, i.e. A, is true for I. So if $\vdash_T A$, then A is true for I.

2. Suppose A is true for I. Then B is, by 40.6. So by the induction hypothesis $\vdash_T B$. So by 45.4 $\vdash_T \Lambda v_j B$, i.e. $\vdash_T A$. So if A is true for I, $\vdash_T A$.

II. B is open. Since A is closed, the only free variable in B is $v_j$ for some *j*.

1. Suppose $\vdash_T A$, i.e. $\vdash_T \Lambda v_j B$. Then by K4 $\vdash_T Bt/v_j$ for each closed term t of T. So by the induction hypothesis $Bt/v_j$ is true for I for each closed term t of T. So by 40.21 $\Lambda v_j B$, i.e. A, is true for I. So if $\vdash_T A$, then A is true for I.

2. Suppose A, i.e. $\Lambda v_j B$, is true for I. Then B is, by 40.6. So by 40.20, $Bt/v_j$ is true for I for every closed term t of T. So by the induction hypothesis $\vdash_T Bt/v_j$ for every closed term t of T. So by the closedness of T $\vdash_T \Lambda v_j B$, i.e. $\vdash_T A$. So if A is true for I, $\vdash_T A$.

*Comment.* In order to show that I is a model of T, all we really want is: If $\vdash_T A$, then A is true for I. We prove the stronger proposition, $\vdash_T A$ iff A is true for I, because we need the 'iff' in the induction hypothesis in order to prove the weaker one. Cf. the similar comment on the proof of 32.13 (above, p. 111).

This proof skirts the edge of vicious circularity without quite falling into it. We want to prove that if a formula of T is a theorem then it is true for a certain interpretation; and we then define the interpretation in question as one in which an atomic formula of T is true iff it is a theorem of T. For an atomic formula to be true, on this interpretation, is just for it to be a theorem of T: this is what we mean by 'true' (for atomic formulas) for the interpretation in question. T is in fact interpreted in terms of itself. This will probably strike the reader as something of a cheat, especially if he was expecting a more interesting model.

45.15 *Any consistent first order theory has a denumerable model*
[The key theorem in Part 3]

*Proof.* Directly from

45.13 If K is a consistent first order theory, then there is a first order theory K′ that is a consistent, closed, and negation-complete extension of K

and

45.14 Any consistent, closed, negation-complete first order theory has a denumerable model

together with the fact that if S′ is an extension of S, then any model of S′ is a model of S.

Alternatively,[1] recapitulating the stages in the proof of 45.13: Let K be an arbitrary consistent first order theory.

1. Enlarge K to $K_0$, by adding a denumerable set of new individual constants. By 45.11 $K_0$ is a consistent first order theory and an extension of K.
2. Enlarge $K_0$ to $K_\infty$, by adding denumerably many proper axioms of the form
$$A_n b_{j_n}/v_n \supset \Lambda v_n A_n$$
By the proof of 45.13 $K_\infty$ is a consistent, *closed* first order theory, and an extension of $K_0$ and therefore of K.
3. Enlarge $K_\infty$ to K′ by adding one by one as axioms each closed wff that can consistently be added at that stage. By 45.10 and the end of the proof of 45.13 K′ is a consistent, closed, *negation-complete* first order theory, and an extension of $K_\infty$, and therefore of K.
4. By 45.14 K′ has a denumerable model, and therefore K, of which K′ is an extension, also has a denumerable model.

45.16 (Alternative version of 45.14) *Any consistent set of closed wffs of a first order theory has a denumerable model*

*Proof.* Let Γ be a consistent set of *closed* wffs of a first order theory K. Then K + Γ is a consistent first order theory, and

[1] This is, in essentials, Henkin's proof [Henkin, 1947].

therefore by 45.15 it has a denumerable model, and therefore Γ has a denumerable model.

We cannot drop the requirement in 45.16 that the wffs be *closed* wffs. If Γ were a set of arbitrary wffs of K, then K + Γ would not necessarily be a first order theory, for all the proper axioms of a first order theory, as we have defined ' first order theory', must be closed wffs; and so we could not apply 45.15 to K + Γ. And in fact there are consistent sets of wffs of first order theories that have no models. E.g. the set

$$\{F^{*\prime}x', \ {\sim}F^{*\prime}x''\}$$

or, as we shall write it,

$$\{Fx, \ {\sim}Fy\}$$

is a consistent set of QS that has no model. *Proof*: Suppose $\{Fx, \ {\sim}Fy\}$ is an inconsistent set of QS. Then Fx, ${\sim}$Fy $\vdash_{QS}$ A, where A is any arbitrary wff of Q. Let A be ${\sim}(p' \supset p')$, which is satisfied by no sequence. Then $Fx$, ${\sim}Fy \vdash_{QS} {\sim}(p' \supset p')$. Then by 43.7 $Fx$, ${\sim}Fy \vDash_{Q} {\sim}(p' \supset p')$. But this last proposition is certainly false: there are sequences that satisfy $\{Fx, \ {\sim}Fy\}$ but do not satisfy ${\sim}(p' \supset p')$. (E.g. Let I be an interpretation of Q whose domain is the set of natural numbers. Let I assign truth to $p'$, and the property of *being an even number* to the predicate symbol $F$ [i.e. $F^{*\prime}$]. Then the sequence ⟨2, 3, 4, 5, . . .⟩ satisfies both $Fx$ and ${\sim}Fy$ [i.e. $Fx'$ and ${\sim}Fx''$] but does not satisfy ${\sim}(p' \supset p')$.) So {Fx, ${\sim}$Fy} is a consistent set of QS. But $\{Fx, {\sim}Fy\}$ has no model. *Proof*: Suppose that there were a model M of $\{Fx, \ {\sim}Fy\}$. Then $Fx$ and ${\sim}Fy$ would both be true for M. So by 40.6 $\bigwedge xFx$ and $\bigwedge y{\sim}Fy$ would both be true for M, and so M would be a model of $\{\bigwedge xFx, \bigwedge y{\sim}Fy\}$. But $\bigwedge xFx \vdash_{QS} Fy$ (using $\vdash_{QS} \bigwedge xFx \supset Fy$, by QS4) and $\bigwedge y{\sim}Fy \vdash_{QS} {\sim}Fy$ (using $\vdash_{QS} \bigwedge y {\sim}Fy \supset {\sim}Fy$, again by QS4). So by 43.7 $\bigwedge xFx \vDash_{Q} Fy$ and $\bigwedge y{\sim}Fy \vDash_{Q} {\sim}Fy$. So any model of $\{\bigwedge xFx, \bigwedge y{\sim}Fy\}$ would have to be a model both of $Fy$ and of ${\sim}Fy$, which is impossible. So {⋀xFx, ⋀y${\sim}$Fy} has no model. So {Fx, ${\sim}$Fy} has no model.

However, we can prove the following:

45.17 *Any consistent set of wffs of a first order theory is simultaneously satisfiable in a denumerable domain* [Gödel, 1930]

To get this result we prove, first, that if Γ is a consistent set of a first order theory then the result of putting new constants for free occurrences of variables in Γ is also a consistent set of a first order theory. Then, making use of the earlier result 45.16, we define a sequence that will satisfy not only every member of the new set but also every member of Γ. We begin with some lemmas.

*Lemma 1. If Γ is a consistent set of a first order theory K, and K′ is a first order theory that results from merely adding denumerably many new individual constants to K, then Γ is a consistent set of K′*

*Proof.* Suppose Γ is a set of wffs of K that is an inconsistent set of K′. Then for some formula A of K′ there are derivations in K′ of A and of ~A from Γ. In these derivations replace any constants not in K by individual variables that do not occur in either derivation. The result will be derivations *in K* from Γ of some formula and its negation. For under this transformation formulas that are logical axioms of K′ will be or become logical axioms of K, formulas that are proper axioms of K′ will be proper axioms of K (for K′ has no proper axioms other than those of K), and applications of Modus Ponens in the original derivations will still be applications of Modus Ponens in the transformed strings. So if Γ (a set of wffs of K) is an inconsistent set of K′ it will be an inconsistent set of K. So if Γ is a consistent set of K it is a consistent set of K′.

*Lemma 2. Let A be a wff in which some variable v has one or more free occurrences. Let c be a constant that does not occur in A or in any proper axiom of K. Then if $\vdash_K$ Ac/v, then $\vdash_K$ A*

*Proof.* Suppose $\vdash_K$ Ac/v. Let $A_1, \ldots, A_n$ be some proof in K of Ac/v (so $A_n$ is Ac/v). Let u be a variable that does not occur in any of $A_1, \ldots, A_n$. Let $A_1', \ldots, A_n'$ be the result of replacing all occurrences of c in $A_1, \ldots, A_n$ by the variable u. Then $A_1', \ldots, A_n'$ is a proof in K of Au/v, by a familiar argument (for each $A_i$, if $A_i$ is a logical axiom of K then $A_i'$ is too, and if $A_i$ is a proper axiom of K then c does not occur in $A_i$ and so $A_i'$ is $A_i$ and therefore a proper axiom of K). So $\vdash_K$ Au/v. So by 45.4 $\vdash_K$ ΛuAu/v. But $\vdash_K$ ΛuAu/v ⊃ A (logical axiom of K, by K4: v is free for u in Au/v). So by MP $\vdash_K$ A.

We now prove that the set obtained by putting new constants for free occurrences of variables in a consistent set of a first order theory is itself a consistent set of a first order theory:

Let $\Gamma$ be an arbitrary consistent set of wffs of an arbitrary first order theory K. Let $v_1, \ldots, v_k, \ldots$ be all the variables that have free occurrences in $\Gamma$. Let K′ be the first order theory that results from adding denumerably many new constants $c_1, c_2, c_3, \ldots$ to K. Let $\Gamma'$ be the set of closed wffs obtained by substituting new constants $c_1, \ldots, c_k, \ldots$ for the free occurrences of the variables $v_1, \ldots, v_k, \ldots$ throughout the members of $\Gamma$, distinct constants being substituted for distinct variables. Then $\Gamma'$ is a consistent set of K′. For suppose $\Gamma'$ were an inconsistent set of K′. Then for some finite subset $\Delta$ of $\Gamma'$ and some formula B of K′ we should have both $\Delta \vdash_{K'} B$ and $\Delta \vdash_{K'} \sim B$. Let the members of $\Delta$ be $A_1, \ldots, A_n$. Then we should have

$$A_1, \ldots, A_n \vdash_{K'} B$$

and

$$A_1, \ldots, A_n \vdash_{K'} \sim B$$

and so by the Deduction Theorem, applied $n$ times,

$$\vdash_{K'} (A_1 \supset (\ldots (A_n \supset B) \ldots))$$

and similarly

$$\vdash_{K'} (A_1 \supset (\ldots (A_n \supset \sim B) \ldots))$$

Then by Lemma 2, applied as often as may be necessary, we should have

$$\vdash_{K'} (A_1' \supset (\ldots (A_n' \supset B') \ldots))$$

and

$$\vdash_{K'} (A_1' \supset (\ldots (A_n' \supset \sim B') \ldots))$$

where each $A_i'$ is the result of replacing every new constant in $A_i$ by its corresponding variable, and similarly for B′. So we should have

$$A_1', \ldots, A_n' \vdash_{K'} B'$$

and

$$A_1', \ldots, A_n' \vdash_{K'} \sim B'$$

Now the set $\{A_1', \ldots, A_n'\}$ is a subset of the original set $\Gamma$. So we should have

$$\Gamma \vdash_{K'} B'$$

and

$$\Gamma \vdash_{K'} \sim B';$$

i.e. Γ would be an inconsistent set of K′. So by Lemma 1 Γ would be an inconsistent set of K. But this contradicts our hypothesis that Γ was a consistent set of K. So Γ′ must be a consistent set of K′.

Since Γ′ is consistent, it will have a denumerable model, by 45.16. Let M be a denumerable model of Γ′. We define a denumerable sequence *s* of members of the domain of M as follows: If $v_1$ is the *m*th variable in the fixed enumeration of the variables in the interpretation M, then *s* is to have for its *m*th term the object assigned by M to the constant $c_1$. Similarly, if $v_2$ is the *n*th variable in the fixed enumeration, then *s* is to have for its *n*th term the object assigned by M to $c_2$. And so on for all of $v_1, \ldots, v_k, \ldots$ For any remaining terms of *s* we may take arbitrary members of the domain of M. Since M is a model of Γ′, *s* satisfies every member of Γ′. But it is clear, from the way we have defined *s*, that *s* will also satisfy every member of Γ. Hence if Γ is consistent, Γ is simultaneously satisfiable in a denumerable domain.

Q.E.D.

45.18 (*The Löwenheim–Skolem Theorem*) *If a first order theory has a model, then it has a denumerable model*

*Proof.* Directly from 45.8 (if a first order theory has a model, then it is consistent) and 45.15 (any consistent first order theory has a denumerable model).

We follow custom and call 45.18 'The Löwenheim–Skolem Theorem' – rather unhistorically, since the relevant theorems of Löwenheim and Skolem belong to pure model theory and make no mention of any deductive apparatus. (Also, our notion of model, which is due to Alfred Tarski, comes from a later time.)

*Historical note*

Leopold Löwenheim (1878–*c.* 1940) published in 1915 the theorem that (in our words, not his):

> If a formula of the predicate calculus with identity is true for every normal interpretation with a finite domain, then if it is true for every normal interpretation with a denumerable domain then it is true for every normal interpretation

where a normal interpretation is one in which the symbol '=' (or some 2-place predicate symbol) is interpreted as meaning 'is identical with'. Löwenheim's proof had a slight gap in it.

Thoralf Skolem (1887–1963) proved in 1919:
If a set of formulas of the predicate calculus is simultaneously satisfiable in any non-empty domain, then it is simultaneously satisfiable in some denumerable domain.

45.19 *If there is an interpretation I for which every proper axiom of a first order theory K is true, then K has a model*

*Proof*. Every logical axiom of K is logically valid, and so true for I. Suppose every proper axiom of K is also true for I. Then every axiom of K is true for I. Modus Ponens preserves truth for I. So every theorem of K is true for I; i.e. K has a model.

45.20 (*The Compactness Theorem: Gödel, 1930, in pure model-theoretic form* (*cf. 32.20*)) *If every finite subset of the set of proper axioms of a first order theory K has a model, then K has a model*

*Proof*. Suppose K has no model. Then K is inconsistent, by 45.15. So, by a familiar argument, there is a finite subset Δ of proper axioms of K such that for some formula A there is a proof in K of A and a proof in K of ~A, and these proofs use no proper axioms that are not in Δ. Take Δ as the set of proper axioms of a new first order theory K′ with the same language as K. Then $\vdash_{K'}$ A and $\vdash_{K'}$ ~A. So K′ is inconsistent. So by 45.8 K′ has no model. So by 45.19 there is no interpretation for which every proper axiom of K′ is true; i.e. Δ has no model; i.e. a finite subset of the set of proper axioms of K has no model. So if every finite subset of the set of proper axioms of K has a model, then K has a model.

The remaining metatheorems in this section will be used only in Part 4. They could be skipped on a first reading of Part 3.

45.21 *If* $\Gamma \vdash_K A$ *and v does not occur free in Γ, then* $\Gamma \vdash_K \bigwedge vA$ (This is an extension of 45.4)

*Proof* (a simple modification and extension of the proof of 45.4). Assume that $\Gamma \vdash_K$ A and that v does not occur free in Γ. We show that $\Gamma \vdash_K \bigwedge$vA, by induction on the length, *n*, of a derivation of A from Γ.

*Basis*: $n = 1$

Three possibilities:

1. A is a logical axiom of K
2. A is a proper axiom of K
3. A is in Γ

1. Suppose A is a logical axiom of K. Then $\Lambda$vA is also a logical axiom of K, by K7. So $\Gamma \vdash_K \Lambda$vA.

2. Suppose A is a proper axiom of K. Then $\Lambda$vA is a theorem of K, by the steps

A
A ⊃ $\Lambda$vA [K5. A, being a proper axiom, is a closed wff]
$\Lambda$vA

So $\Gamma \vdash_K \Lambda$vA.

3. Suppose A is in Γ. Then $\Lambda$vA is derivable from Γ, by the steps

A
A ⊃ $\Lambda$vA [K5. By hypothesis v is not free in Γ and therefore not free in A]
$\Lambda$vA

So $\Gamma \vdash_K \Lambda$vA.

*Induction Step*

Assume the theorem holds for all derivations (of A from Γ) of length less than $k$. To prove it holds for all derivations of length $k$. 4 cases:

1–3. As in the Basis
4. A is an immediate consequence by MP of two preceding wffs

*Cases* 1–3. As in the Basis.

*Case* 4. Let $A_1, \ldots, A_k$ be a derivation of A from Γ (so $A_k$ is A). A is an immediate consequence by MP of $A_i$ and $A_j$ where $i<k$ and $j<k$ and $A_j$ is $A_i \supset A$. Then by the induction hypothesis $\Gamma \vdash_K \Lambda vA_i$ and $\Gamma \vdash_K \Lambda v(A_i \supset A)$. By K6 $\vdash_K \Lambda v(A_i \supset A) \supset (\Lambda vA_i \supset \Lambda vA)$. So by MP twice $\Gamma \vdash_K \Lambda vA$.

45.22 *If v has no free occurrence in B, then* $\vdash_K \Lambda v(A \supset B) \supset (\bigvee vA \supset B)$

*Proof.* Abbreviate $\Lambda$v(A ⊃ B) to P. Let v have no free occurrence in B. Then:

1. $P \vdash_K \Lambda v(A \supset B)$ [P is $\Lambda$v(A ⊃ B)]
2. $\vdash_K \Lambda v(A \supset B) \supset (A \supset B)$ K4

3. $P \vdash_K A \supset B$ — 1, 2, MP
4. $P \vdash_K \sim B \supset \sim A$ — From 3 by propositional logic [i.e. using only instances of tautological schemata, and MP]
5. $P \vdash_K \bigwedge v(\sim B \supset \sim A)$ — From 4 by 45.21 [v does not occur free in P]
6. $\vdash_K \bigwedge v(\sim B \supset \sim A) \supset (\bigwedge v \sim B \supset \bigwedge v \sim A)$ — K6
7. $P \vdash_K \bigwedge v \sim B \supset \bigwedge v \sim A$ — 5, 6, MP
8. $\vdash_K \sim B \supset \bigwedge v \sim B$ — K5 [v does not occur free in B]
9. $P \vdash_K \sim B \supset \bigwedge v \sim A$ — 7, 8, propositional logic
10. $P \vdash_K \sim \bigwedge v \sim A \supset B$ — 9, propositional logic
    i.e. $\bigwedge v(A \supset B) \vdash_K \bigvee vA \supset B$

So by the Deduction Theorem $\vdash_K \bigwedge v(A \supset B) \supset (\bigvee vA \supset B)$

45.23 *If the variable u does not occur in A, then* $\vdash_K \bigvee vA \supset \bigvee uAu/v$

*Proof.* Let u be a variable that does not occur in A. Then:

1. $\vdash_K \bigwedge u \sim Au/v \supset \sim A$ — K4 [Since u does not occur in $\sim A$, and $\sim Au/v$ is just the result of replacing all free occurrences of v in $\sim A$ by u, wherever v is free in $\sim A$ u will be free in $\sim Au/v$, and conversely]
2. $\vdash_K A \supset \sim \bigwedge u \sim Au/v$ — 1, propositional logic
   i.e. $\vdash_K A \supset \bigvee uAu/v$
3. $\vdash_K \bigwedge v(A \supset \bigvee uAu/v)$ — 2, 45.4
4. $\vdash_K \bigwedge v(A \supset \bigvee uAu/v) \supset (\bigvee vA \supset \bigvee uAu/v)$ — 45.22 [v has no free occurrence in $\bigvee uAu/v$]
5. $\vdash_K \bigvee vA \supset \bigvee uAu/v$ — 3, 4, MP

45.24 *Let K be an arbitrary first order theory. Let K′ be a first order theory that results from merely adding denumerably many new individual constants to K. Let A be a formula of K in which a variable v has one or more free occurrences. Then if* $\Gamma \vdash_K \bigvee vA$ *and* $\Gamma$, $Ac/v \vdash_{K'} B$, *where c is one of the constants in K′ but not in K, and c does not occur in B, then* $\Gamma \vdash_K B$

45.24, in one version or another, is often called *Rule C* [The Rule of Choice], the name being due to Rosser (1953, p. 128: cf.

also Mendelson, p. 73, Margaris, p. 79). It is called the Rule of Choice because it is a formal analogue of a method of argument found in everyday mathematics in which something that might be called an act of choice occurs. Roughly, what happens in the mathematical arguments is that the mathematician, having shown that a certain property does belong to something or other, says 'Let N be a thing that has this property', where 'N' is an arbitrarily chosen name and nothing is to be assumed about N except that it has the property in question. He then continues with his proof and ends with a theorem in which the object N is not mentioned. Notice, however, that 45.24 itself is a metatheorem of pure proof theory and says nothing about acts of choice of any kind.

*Proof.* Suppose $\Gamma \vdash_K \bigvee vA$ and also $\Gamma, Ac/v \vdash_{K'} B$, where c is one of the constants in K′ but not in K, and c does not occur in B. By the Deduction Theorem $\Gamma \vdash_{K'} Ac/v \supset B$. There will therefore be a derivation in K′ of Ac/v from a finite subset Δ of Γ. In this derivation replace any distinct constants $c_1, \ldots, c_n$ that are not in K by distinct variables $u_1, \ldots, u_n$ that do not occur in the derivation. Now K′ has no proper axioms other than those of K; and under the transformation any logical axioms of K′ in the original derivation become logical axioms of K, and applications of MP in the original derivation are still applications of MP. So the transformation yields a derivation in K of $Au/v \supset B$ from Δ, where u is one of the variables in the set $\{u_1, \ldots, u_n\}$. So $\Delta \vdash_K Au/v \supset B$, where u does not occur in Δ or in A or in B.

So 1. $\Delta \vdash_K \bigwedge u(Au/v \supset B)$, by 45.21.
So 2. $\Delta \vdash_K \bigvee uAu/v \supset B$, by 1, 45.22 and MP.
So 3. $\Gamma \vdash_K \bigvee uAu/v \supset B$, since Δ is a subset of Γ.
Now by hypothesis
4. $\Gamma \vdash_K \bigvee vA$
So 5. $\Gamma \vdash_K \bigvee uAu/v$, by 4, 45.23 and MP.
So 6. $\Gamma \vdash_K B$, from 3 and 5 by MP.

45.25 *If t is a closed term of K, then* $\vdash_K At/v \supset \bigvee vA$

*Proof*

1. $\vdash_K \bigwedge v{\sim}A \supset {\sim}At/v$ K4 [Since t is a closed term, t is free for v in A]

2. $\vdash_K At/v \supset \sim\Lambda v \sim A$ From 1 by propositional logic
i.e. $\vdash_K At/v \supset \vee vA$

*Note on the definition of 'first order theory'*

Two defining features of first order theories, as we use the term, are these:

(1) Modus Ponens for $\supset$ is their *sole* rule of inference.
(2) All their proper axioms are *closed* wffs.

But other definitions of 'first order theory' may be found in the literature. For example, Elliott Mendelson's definition (1964) is very different from ours. First order theories, in his sense, have an extra rule of inference, viz. the Rule of Generalization:

$\Lambda v_k A$ is an immediate consequence of A

Mendelson also allows open wffs to be proper axioms of his first order theories. In consequence, for his first order theories the Deduction Theorem holds only in a restricted form, and neither

$$\text{If } \Gamma \vdash A \text{ then } \Gamma \vDash A$$

nor

$$\text{If } A \vdash B \text{ then } \vDash A \supset B$$

holds for them.

Another consequence of Mendelson's definition of 'first order theory' is that some sets that are consistent by our definition are inconsistent by his: e.g. the set $\{Fx, \sim Fy\}$. Consequently he is able to establish

Any consistent set of wffs of a first order theory [his sense] has a denumerable model

while we can only get

Any consistent set of wffs of a first order theory [our sense] is simultaneously satisfiable in some denumerable domain (45.17)

However we can get

Any consistent first order theory has a denumerable model (45.15)

and also it is true of our[1] system of first order predicate logic, but not of his, that syntactic consequence and semantic consequence coincide, i.e. $\Gamma \vdash_{QS} A$ iff $\Gamma \vDash_Q A$ [46.3]

[1] There is nothing proprietorial about this 'our'. The essential idea is due to Frederic B. Fitch (1938, pp. 144–5), and Fitch-type systems may be found in Quine (1940), Rosser (1953), Mates (1965), and Margaris (1967).

## 46 Proof of the semantic completeness of QS

46.1 (*The semantic completeness theorem for QS*) *Every logically valid formula of Q is a theorem of QS*: *i.e. If* $\vDash_Q A$ *then* $\vdash_{QS} A$

*Proof*

1. Suppose that C is a *closed* wff of Q that is not a theorem of QS. Then QS +{~C} is a consistent first order theory, by 45.6. So QS +{~C} has a model, by 45.15. So ~C has a model. So C is not logically valid. So if C is a logically valid closed wff of Q, then C is a theorem of QS.

2. Suppose that A is a wff (open or closed) of Q that is logically valid. Then an arbitrary closure of A, $A^c$, is logically valid, by 40.8. So $A^c$ is a theorem of QS, by Step 1 above. So A is a theorem of QS, by 45.5. So if A is a logically valid wff of Q, then A is a theorem of QS.

*Alternative proof*

1. Suppose that C is a closed wff of Q that is not a theorem of QS. Then {~C} is a consistent set [of closed wffs] of QS, by 45.7. So {~C} has a model, by 45.16. The rest of the proof is as above.

46.2 (*The 'strong' completeness theorem for QS*) *If* $\Gamma \vDash_Q A$, *then* $\Gamma \vdash_{QS} A$

*Proof.* Suppose $\Gamma \vDash_Q$ A. Then $\Gamma \cup \{\sim A\}$ is not simultaneously satisfiable. So $\Gamma \cup \{\sim A\}$ is an inconsistent set of QS, by 45.17. So $\Gamma \vdash_{QS}$ A, by 43.9. So if $\Gamma \vDash_Q$ A, then $\Gamma \vdash_{QS}$ A.

46.3 $\Gamma \vdash_{QS} A$ *iff* $\Gamma \vDash_Q A$

*Proof.* Directly from 46.2 and 43.7 (if $\Gamma \vdash_{QS}$ A, then $\Gamma \vDash_Q$ A).

QS is neither negation-complete nor syntactically complete:

46.4 *QS is not negation-complete*

*Proof.* Neither $\Lambda x'F^{*\prime}x'$ nor $\sim\Lambda x'F^{*\prime}x'$ is a theorem of QS, for neither is logically valid, and all theorems of QS are logically valid. [We could just as well have taken $p'$ and $\sim p'$.]

46.5 *QS is not syntactically complete*

*Proof.* The schema

$$C \supset \Lambda x' F^{*\prime} x'$$

where C is an arbitrary closed wff is not provable in QS (for if it were, then $\Lambda x' F^{*\prime} x'$ would be a theorem of QS, taking some closed wff that is an axiom as C and then using Modus Ponens: and we have just seen – 46.4 – that it is not). We shall show that this schema can consistently be added to QS as an axiom-schema:

Let I be an interpretation of Q whose domain, D, is the set of all closed terms of Q. Let I assign to $F^{*\prime}$ the set of all closed terms of Q. [So, intuitively, '$F^{*\prime}$' is interpreted as meaning 'is a closed term of Q'.] Then every denumerable sequence of members of D satisfies $F^{*\prime} x'$. So $\Lambda x' F^{*\prime} x'$ is true for I. So, for any closed wff C, $C \supset \Lambda x' F^{*\prime} x'$ is true for I. Let S be the formal system that results from adding $C \supset \Lambda x' F^{*\prime} x'$ as an axiom-schema to QS (C being restricted to closed wffs). Then every axiom of S is true for I (the axioms of QS are true for I, because they are all true for every interpretation). Modus Ponens preserves truth for an interpretation. So the theorems of S are all true for I. So S is consistent.

So there is an unprovable schema that can consistently be added to QS as an axiom-schema. So QS is not syntactically complete.

## 47 A formal system of first order predicate logic with identity: the system QS=. Proof of the consistency of QS=. Normal models. Proof of the adequacy of QS=

QS= is a first order theory with the same language as QS (i.e. the language Q) and denumerably many proper axioms, viz. the axiom

QS= 1 $\Lambda x' F^{**\prime} x' x'$

and the axioms given by QS= 2:

QS= 2 Every closure of $F^{**\prime} x' x'' \supset (A \supset A')$, where A and A′ are wffs of Q and A′ is like A except that $x''$ may replace any free occurrence of $x'$ in A, provided that

$x''$ occurs free wherever it replaces $x'$. [$x''$ need not be substituted for *every* free occurrence of $x'$ in A.]

If we write $=$ for $F^{**'}$, $x$ for $x'$, $y$ for $x''$, and $x' = x''$ for $=x'x''$, and add square brackets to indicate grouping, the intended interpretation of the new axioms becomes clearer:

QS= 1 $\Lambda x[x = x]$

QS= 2 Every closure of $x = y \supset (A \supset A')$, where A and A′ are as explained.

*Consistency of QS=*

For each wff A of Q we define its associated propositional formula (a.p.f.), $A^{prop}$, as follows:

Delete from A all quantifiers and their attendant variables.[1] Replace each term by $x'$. In the resulting string replace each occurrence of $F^{**'}x'x'$ by $(p' \supset p')$. In the resulting string delete all terms and replace each predicate symbol by $p'$. The result, $A^{prop}$, is a formula of P.

We shall call this procedure the *transformation* of A into $A^{prop}$.

We show that the a.p.f. of each axiom of QS= is a tautology of P. Anything that is an immediate consequence in QS= by Modus Ponens of two formulas each of which has a tautology of P as its a.p.f. will also have a tautology of P as its a.p.f.: for $(A \supset B)^{prop}$ is $(A^{prop} \supset B^{prop})$. So each theorem of QS= has a tautology of P as its a.p.f. But the formula

$$\sim F^{**'}x'x'$$

of Q has $\sim(p' \supset p')$ as its a.p.f., and this is not a tautology of P. So $\sim F^{**'}x'x'$ is not a theorem of QS=. So QS= is consistent. It remains to prove that each axiom of QS= has a tautology of P as its a.p.f.

It is obvious that anything that is an axiom by any of QS 1–3 will have a tautology of P as its a.p.f.

Anything that is an axiom of QS= by QS 4 [$\Lambda vA \supset At/v$, if t is free for v in A] becomes at the first stage of the transformation $A' \supset A't/v$ where A′ has no quantifiers and $A't/v$ is like A′ except that the term t may replace the variable v in some of its occurrences in A′. At the next stage we have $A'' \supset A''$ where A″ results from A′ or $A't/v$ by replacing each term in A′ or

[1] i.e. the variables *immediately* following the quantifiers.

A′t/v by $x'$. From then on any changes affect both sides of the $\supset$ equally, and the final result is the tautology $A^{prop} \supset A^{prop}$.

Anything that is an axiom by QS 5 [$A \supset \wedge vA$, if v does not occur free in A] becomes at the first stage $A' \supset A'$ where A′ has no quantifiers, and finally $A^{prop} \supset A^{prop}$, a tautology of P.

Anything that is an axiom by QS 6 [$\wedge v(A \supset B) \supset (\wedge vA \supset \wedge vB)$] becomes first $(A' \supset B') \supset (A' \supset B')$ where A′ and B′ have no quantifiers, and finally the tautology of P, $(A^{prop} \supset B^{prop}) \supset (A^{prop} \supset B^{prop})$.

In view of what has gone before, anything that is an axiom by QS 7 will have a tautology of P as its a.p.f.

$QS^{=}$ 1 [$\wedge x' F^{**\prime} x' x'$] becomes first $F^{**\prime} x' x'$ and then $(p' \supset p')$ a tautology of P.

Anything that is an axiom by $QS^{=}$ 2 [every closure of $F^{**\prime} x' x'' \supset (A \supset A')$] becomes at the first stage $F^{**\prime} x' x'' \supset (A^{*} \supset A'^{*})$ where $A^{*}$ and $A'^{*}$ have no quantifiers. Then it becomes $F^{**\prime} x' x' \supset (A^{*R} \supset A^{*R})$ where $A^{*R}$ is like $A^{*}$ except that every term in $A^{*}$ is replaced by $x'$. Finally we get $(p' \supset p') \supset (A^{prop} \supset A^{prop})$, a tautology of P.

So:

47.1 *$QS^{=}$ is consistent*

47.4 gives us another proof of consistency.

*Definition.* I is a *normal interpretation* of Q ($Q^{+}$) iff I is an interpretation of Q ($Q^{+}$) and I assigns the relation of identity, defined for members of the domain of I, to the predicate symbol $F^{**\prime}$.

We shall say that a first order theory K is a *first order theory with identity* if K has $QS^{=}$ 1 as an axiom and $QS^{=}$ 2 (applied to wffs of the language of K) as an axiom-schema.

*Definition.* I is a *normal model* of a first order theory K with identity iff I is a normal interpretation of the language of K and I is a model of K.

47.2 *If K is a consistent first order theory with identity, then K has a countable normal model*

*Outline of proof.* Let K be a consistent first order theory with identity. By the proof of 45.15 K has a denumerable model, M, whose domain, D, is the set of all closed terms of K. Let the closed terms of K be distributed into disjoint sets (i.e. sets with no members in common) $S_1$, $S_2$, etc. as follows:

Let $c_{j1}$ be the first closed term in some enumeration of the closed terms of K. Then $c_{j1}$ is to be a member of $S_1$, and, for any closed term t, t is to be a member of $S_1$ iff $\vdash_K t = c_{j1}$ [i.e. $\vdash_K F^{**\prime} t c_{j1}$]. Let $c_{j2}$ be the first closed term (if any) in the enumeration that is not a member of $S_1$. Then $c_{j2}$ is to be a member of $S_2$, and, for any closed term t, t is to be a member of $S_2$ iff $\vdash_K t = c_{j2}$. And so on.

Each closed term of K belongs to one and only one $S_i$, and for each closed term t $\vdash_K t = c_{ji}$ for some positive integer $i$ (proofs left as exercises: $\vdash_K t = t$ for each term t in K, by QS= 1 and K4.)

Let M′ be the interpretation of K whose domain D′ is the set $\{c_{j1}, c_{j2}, c_{j3}, \ldots\}$. This set is countable, but not necessarily denumerable: e.g. if for every closed term t $\vdash_K t = c_{j1}$, then D′ is $\{c_{j1}\}$. To each closed term t in K M′ assigns the closed term $c_{ji}$ in D′ such that $\vdash_K t = c_{ji}$. To $F^{**\prime}$ M′ assigns the relation of identity (for the domain D′). In all other respects M′ is like M.

Let $s = \langle d_1, d_2, \ldots \rangle$ be a denumerable sequence of members of D, and let $s' = \langle d_1', d_2', \ldots \rangle$ be the denumerable sequence of members of D′ that results from $s$ when each $d_i$ in $s$ is replaced by the member $d_i'$ of D′ such that $\vdash_K d_i = d_i'$ ($d_i$ and $d_i'$ are closed terms of K). Then it can be proved by induction on the number of quantifiers and connectives in A that a wff A is satisfied by $s$ in M iff A is satisfied by $s'$ in M′. It follows that, for any wff A of K, A is true for M iff A is true for M′. Since M is a model of K, M′ is a model of K. It is also a normal model of K, and it has a countable domain.

*Adequacy of QS=*

We want QS=

(1) to have as a theorem every logically valid wff of Q and in addition every wff of Q that, though not logically valid, is true for every interpretation of Q for which $F^{**\prime}$ is interpreted as meaning 'is identical with';

(2) to have as theorems *only* such wffs.

That is:

1. We want any formula of Q that is true for every normal interpretation of Q to be a theorem of QS=.

2. We want every theorem of $QS^=$ to be true for every normal interpretation of Q.

47.3 *Any formula of Q that is true for every normal interpretation of Q is a theorem of $QS^=$*

*Proof*

1. Suppose C is a closed wff of Q that is true for every normal interpretation of Q. Let $QS^= + \{\sim C\}$ be the result of adding $\sim C$ to $QS^=$ as a proper axiom. If M is any normal model for $QS^= + \{\sim C\}$, then both C and $\sim C$ are true for M. This is impossible. So $QS^= + \{\sim C\}$ has no normal model. But $QS^= + \{\sim C\}$ is a first-order theory with identity, so by 47.2 it is inconsistent. So, by 45.6, C is a theorem of $QS^=$.

2. Let A be a wff (open or closed) of Q that is true for every normal interpretation of Q. Then by 40.7 $A^c$ is true for every normal interpretation of Q. So $A^c$ is a theorem of $QS^=$, by Step 1 above. So A is a theorem of $QS^=$, by 45.5.

47.4 *Every theorem of $QS^=$ is true for every normal interpretation of Q*

*Proof.* The logical axioms of $QS^=$ are the axioms of QS. Every axiom of QS is logically valid (cf. 43.5) and therefore true for every interpretation and therefore true for every normal interpretation.

The proper axioms of $QS^=$ are $QS^=$ 1 and everything that is allowed as an axiom by $QS^=$ 2.

$QS^=$ 1 is true for every normal interpretation of Q.

$QS^=$ 2 allows as an axiom any closure of $F^{**\prime}x'x'' \supset (A \supset A')$, where A′ is like A except that $x''$ may replace any free occurrence of $x'$ in A, provided that $x''$ occurs free wherever it replaces $x'$. Let N be any normal interpretation of Q. A denumerable sequence $s$ satisfies the left-hand side of $QS^=$ 2 for N iff its first two terms are the same. Since A′ differs from A at most in having some free occurrence(s) of $x''$ where A has some free occurrence(s) of $x'$, any sequence whose first two terms are the same and that satisfies A for N will satisfy A′ for N. So any sequence that satisfies the left-hand side for N will satisfy the right-hand side for N. So every sequence of a normal interpretation satisfies $F^{**\prime}x'x'' \supset (A \supset A')$. So every sequence of a normal interpretation satisfies any closure of that expression, and therefore any axiom by $QS^{=}2$. So any axiom by $QS^{=}2$ is true for every normal interpretation.

So every axiom of $QS^{=}$ is true for every normal interpretation. Modus Ponens for $\supset$ preserves truth for an interpretation and therefore truth for a normal interpretation. So every theorem of $QS^{=}$ is true for every normal interpretation.

47.4 gives us another proof of the consistency of $QS^{=}$: for $\sim F^{**\prime}x'x'$ (for example) is not true for any normal interpretation of Q and therefore it is not a theorem of $QS^{=}$.

## 48 Isomorphism of models. Categoricity. Non-standard models

For this section we shall re-define 'first order theory' to mean 'formal system that would be a first order theory by our old definition but for the fact that *it has no propositional symbols*'. All metatheorems proved up to this point for first order theories in our old sense hold equally for first order theories in our new sense. In particular:

1. Any consistent first order theory has a denumerable model.
2. (The Löwenheim–Skolem Theorem) If a first order theory has a model it has a denumerable model.

Further, Tarski has shown (1928) (for first order theories in either sense) that:

3. (The Upward Löwenheim–Skolem Theorem) If a first order theory has an infinite model, then it has a model of any arbitrary infinite cardinality; and if a first order theory has an infinite normal model, then it has a normal model of any arbitrary infinite cardinality.

*Isomorphism of models*

Intuitively speaking, two models of a formal system are isomorphic if they have exactly the same form and differ at most in their content. For first order theories in our new sense (i.e. without propositional symbols) isomorphism comes to this:

Let K be a first order theory. Let M and M′ be models of K, M with domain D, M′ with domain D′. For each constant c of K, let d be the member of D assigned to it by M, d′ the member of D′ assigned to it by M′. For each function symbol f of K, let $f$ be the function assigned to it by M, $f'$ the function assigned to it by M′. For each predicate symbol F, let R be the relation assigned to it by M, R′ the relation assigned to it by M′. Then M is *isomorphic* to M′ iff there is a 1–1 correspondence between D and D′ pairing off each member d of D with a member d′ of D′ in such a way that:

M assigns d to a constant c iff M′ assigns d′ to c.

$f(d_1, \ldots, d_n) = d$ iff $f'(d_1', \ldots, d_n') = d'$.

$d_1, \ldots, d_n$ (in that order) stand in the relation R iff
$d_1', \ldots, d_n'$ (in that order) stand in the relation R′.

It follows from the definition that two models of a first order theory cannot be isomorphic if their domains are of different cardinalities.

We state without proof that

[48.2] *If M and M′ are isomorphic models of a first order theory K, then a formula A of K is true for M iff it is true for M′*

*Categoricity*

The word ‘categorical’ is used in various ways in the literature:

1. A formal system is said to be categorical iff all its models are isomorphic.

In this sense of ‘categorical’ no consistent first order theory is categorical: for by 1 and 3 above any consistent first order theory has models of every infinite cardinality.

2. A first order theory with identity is said to be categorical iff all its *normal* models are isomorphic.

In this sense of ‘categorical’ no first order theory that has an infinite normal model is categorical (by the Upward Löwenheim–Skolem Theorem). But some first order theories that have only finite normal models are categorical: e.g. the first order theory that results from adding

$$\Lambda x \Lambda y[x = y]$$

as a proper axiom to $QS^=$ has only finite normal models (any normal model for it must have a domain with only one member), and all its normal models are isomorphic.

3. A first order theory is said to be $\alpha$-categorical [alpha-

categorical], where $\alpha$ is a cardinal number, iff it has a normal model of cardinality $\alpha$ and any two normal models of cardinality $\alpha$ are isomorphic.

The first order theory mentioned in 2 above is 1-categorical.

48.3 *If all normal models of a first order theory with identity are isomorphic, then the theory is negation-complete* (*i.e. if a first order theory with identity is categorical, in the second of the senses listed above, then it is negation-complete*)

*Proof.* Let K be a first order theory with identity that is not negation-complete. Then for some closed wff C neither $\vdash_K C$ nor $\vdash_K \sim C$. Therefore by 45.6 both K $+\{\sim C\}$ and K $+\{C\}$ are consistent first order theories [with identity]. Therefore both have normal models, by 47.2. Let M be any normal model of K $+\{\sim C\}$ and M′ any normal model of K $+\{C\}$. Then C is false in M and true in M′. So M and M′ are not isomorphic, by [48.2]. So if all normal models of K are isomorphic, K is negation-complete.

*Non-standard models*

A model of a formal system is a *non-standard model* if it is not isomorphic to the standard (or intended) model of the system.

To illustrate the notion of non-standard model, we begin by presenting a set of axioms[1] for a part of arithmetic ('S$x$' on the intended interpretation means 'the successor of $x$'):

1 $\bigwedge x \bigwedge y(Sx = Sy \supset x = y)$
2 $\bigwedge x[0 \neq Sx]$
3 $\bigwedge x(x \neq 0 \supset \bigvee y[x = Sy])$
4 $\bigwedge x[x + 0 = x]$
5 $\bigwedge x \bigwedge y[x + Sy = S[x + y]]$
6 $\bigwedge x[x.0 = 0]$
7 $\bigwedge x \bigwedge y[x.Sy = [x.y] + x]$

These axioms can be formulated in the language of Q, writing

| | |
|---|---|
| $a'$ | for 0 |
| $x'$ | for $x$ |
| $x''$ | for $y$ |
| $f^{*\prime}$ | for S |

[1] This set of axioms is due to R. M. Robinson (1950) and is sometimes called 'Robinson's arithmetic', or 'Q'.

| | |
|---|---|
| $f^{**\prime}\mathrm{t}_1\mathrm{t}_2$ | for $\mathrm{t}_1 + \mathrm{t}_2$, where $\mathrm{t}_1$ and $\mathrm{t}_2$ are terms of Q |
| $f^{**\prime\prime}\mathrm{t}_1\mathrm{t}_2$ | for $\mathrm{t}_1 . \mathrm{t}_2$ |
| $F^{**\prime}\mathrm{t}_1\mathrm{t}_2$ | for $\mathrm{t}_1 = \mathrm{t}_2$ |
| $\sim F^{**\prime}\mathrm{t}_1\mathrm{t}_2$ | for $\mathrm{t}_1 \neq \mathrm{t}_2$ |
| $\sim \Lambda \mathrm{v} \sim$ | for $\mathrm{V}\mathrm{v}$, where v is a variable of Q |

and deleting square brackets. For example, axiom 3 becomes

$$\Lambda x'(\sim F^{**\prime}x'a' \supset \sim \Lambda x'' \sim F^{**\prime}x'f^{*\prime}x'')$$

So re-written the axioms become closed wffs of Q, and if we add them as axioms to QS= we get a first order theory, which we shall call 'R'.

We shall take it for granted that R does have its intended model, M, whose domain is the set of natural numbers; and we shall show that it also has a [normal] model that is not isomorphic to its intended model, i.e. that it has a non-standard [normal] model.

For ease of reading we shall use the usual mathematical symbolism instead of the language Q, writing the numeral 1 instead of $f^{*\prime}a'$, the numeral 2 instead of $f^{*\prime}f^{*\prime}a'$, and so on: we shall also write $c$ instead of $a''$.

Let R′ be the first order theory that has for its proper axioms all the proper axioms of R and also the following formulas of Q (one for each natural number):

$$c \neq 0$$
$$c \neq 1$$
$$c \neq 2$$
$$c \neq 3$$
$$\ldots\ldots$$
$$\ldots\ldots$$

Every finite subset of these new axioms has a model. E.g. every member of the subset $\{c \neq 0, c \neq 1, c \neq 2\}$ comes out true if '$c$' is taken as denoting the number 3. More generally [assuming that R has its intended model], every finite subset of the set of proper axioms of R′ has a model. Therefore by the Compactness Theorem (45.20) R′ has a model. Since R′ is a first order theory with identity (and consistent, since it has a model), by 47.2 it has a normal model, which we shall call M′.

Since M [the standard model] is an interpretation of R and $c$ is a constant of R, M must assign to $c$ some member of its domain D, i.e. some natural number; let us say the number 49. But the number 49 is also assigned by M to the term in Q that we for brevity write as the numeral '49'. So the formula $c = 49$ is true for M. But the formula $c \neq 49$ is an axiom of R′, and therefore true for M′, and therefore $c = 49$ is false for M′. So, by [48.2],[1] M and M′ are not isomorphic. But M′ is a model of R′ and therefore of R. So R has a non-standard model.

We took R merely as an example. It is clear that a similar trick can be played on *any* first order theory with identity that purports to be an axiomatisation of (any part of) arithmetic: if the theory has its intended model, then it also has a normal model that is not isomorphic to its intended model. So:

48.4 *If a first order theory of arithmetic* (*with identity*) *has its intended model, then it also has a normal model that is not isomorphic to its intended model*

This result is due to Skolem (see e.g. his papers of 1933 and 1934). Henkin (1950) has a very simple presentation of the argument.

The theory of non-standard models of arithmetic is known as *non-standard arithmetic*. Abraham Robinson, by using a similar trick on a formal system of the theory of real numbers, has recently developed a theory of *non-standard analysis*,[2] which will be consistent if the theory of real numbers is. In so doing he has made infinitesimals respectable.

## 49 Philosophical implications of some of these results

49.1 *No first order theory can have as its only model a model whose domain is the set of natural numbers*

*Proof.* Every such theory that has a model at all is consistent. So, by the proof of the key lemma in the completeness proof, it has a model whose domain is the set of closed terms of the theory. Closed terms are symbols, not numbers.

[1] If M and M′ are isomorphic models of a first order theory K, then a formula A of K is true for M iff it is true for M′.

[2] *Arithmetic* is the branch of mathematics that deals with the natural numbers and other countable sets of objects. *Analysis* deals with the real numbers and uncountable sets.

49.2 *Let K be any first order theory. We fix the sense of the connectives and quantifiers as usual. We leave the sense of other expressions of K to be determined (so far as this is possible) by the axioms of K. Then even if K has denumerably many axioms they cannot force us to interpret any predicate symbol in K as meaning 'is a natural number' and they cannot force us to interpret any expression in K as the name of a natural number*

*Proof.* If K has a model at all, it has a model whose domain is a set of closed terms, not numbers.

49.3 *As for 49.2 with, in addition, F**' interpreted as meaning identity in the domain of the model*

*Proof.* If K has a model at all, it has a normal model whose domain is a set of closed terms, not numbers.

49.4 *Any first order theory (with identity) that is intended to be an axiomatisation of number theory, if it has a model at all, has a model that is not even isomorphic to its intended model*

This is simply 48.4 in other words.

Nevertheless:

49.5 *The (normal) meaning of any individual numeral adjective that would fill the blank in a sentence of the form 'There are exactly . . . things that have the property F' can be defined in terms of the language of Q, if the quantifiers and connectives are interpreted as usual and some two-place predicate symbol of Q is interpreted as meaning '='*

*Proof.* We show first how to translate 'There is exactly one thing that has the property F' and 'There are exactly two things that have the property F' into the language Q, interpreted as in the statement of the theorem. We write the formulas of Q in abbreviated notation, and then indicate how the abbreviations are to be expanded.

A. There is exactly one thing that has the property F

$$\bigvee x_1(Fx_1 \wedge \bigwedge x_2(Fx_2 \supset x_2 = x_1))$$

B. There are exactly two things that have the property F

$$\bigvee x_1 \bigvee x_2((((Fx_1 \wedge Fx_2) \wedge x_1 \neq x_2) \wedge \bigwedge x_3(Fx_3 \supset (x_3 = x_1 \vee x_3 = x_2))))$$

Expansions:

| | |
|---|---|
| $\bigvee v$ | $\sim\bigwedge v\sim$ |
| $x_1$ | $x'$ |
| $x_2$ | $x''$ |
| $x_3$ | $x'''$ |
| $F$ | Any one-place predicate symbol of Q |
| $(A \wedge B)$ | $\sim(A \supset \sim B)$ |
| $(A \vee B)$ | $(\sim A \supset B)$ |
| $u = v$ | $F^{**\prime}uv$ |
| $u \neq v$ | $\sim F^{**\prime}uv$ |

For 'There are exactly *n* things that have the property F', where *n* is a numeral adjective, and the natural number corresponding to it is greater than 2, we write:

$$\bigvee x_1 \ldots \bigvee x_n(Fx_1 \wedge \ldots \wedge Fx_n \wedge x_1 \neq x_2 \wedge \ldots \wedge x_1 \neq x_n \wedge x_2 \neq x_3 \wedge \ldots \wedge x_2 \neq x_n \wedge \ldots \wedge x_{n-1} \neq x_n \wedge \bigwedge x_{n+1}(Fx_{n+1} \supset (x_{n+1} = x_1 \vee x_{n+1} = x_2 \vee \ldots \vee x_{n+1} = x_n)))$$

filling in the gaps and adding brackets as required.

Assuming that the (normal) meaning of numeral adjectives in other sentences and phrases, including interrogatives and imperatives, is the same as their meaning in sentences of the form 'There are exactly . . . things that have the property F', we get the rather surprising result that:

[49.6] *Any numeral adjective, but no numeral noun, can be unambiguously defined in the language Q, interpreted only in respect of the quantifiers, the connectives, and a symbol for identity*

*The Skolem Paradox*

Set theory can be formulated as a first order theory. For the $\in$ of set-membership we choose some two-place predicate symbol of Q, and then add to QS or $QS^=$ axioms that on their intended interpretation are axioms of set theory. It is not certain that any of the usual axiomatisations of set theory is consistent. So it is not certain that any first order formalisation of set theory has its intended model. But if it has, then the domain of the model is the set of all sets dealt with by the theory. One of the theorems of standard set theory is that there is a set that has uncountably many sets as members (e.g. the set of all subsets of the set of natural numbers is such a set). So the intended model must have an uncountable domain. But by the Löwenheim–Skolem Theorem (45.18) any first order theory that has a model has a *denumerable* model (and, if it is a first order theory with identity, a countable *normal* model).

So the so-called Skolem Paradox is: There is a first order theory that if it has its intended model has an uncountable model. But also if it has its intended model it has a countable model, and even one in which '=' means '='.

The Skolem Paradox is not a contradiction, and by this stage of the book it is hardly even a paradox. Let K be a first order axiomatisation of set theory, and let U be a theorem of K that on its intended interpretation means 'There are uncountably many sets'. Then if K has a model, there is a model of K in which U does not say that there are uncountably many sets, or even that there are uncountably many things of any sort. As our proofs of the relevant metatheorems show, if K has a model, then one normal model of K will make U speak, not of sets at all, but of closed terms of K. Since in that model U will be true, and there are only countably many closed terms of K, U will not mean in that model 'There are uncountably many closed terms': rather it will express some truth about the closed terms of K. And if there is a model of K, with a countable domain, in which U does speak about sets, then still in the model U will not mean 'There are uncountably many sets of such and such a sort' but something different.

## 50 A formal system of first order monadic predicate logic: the system $QS^M$. Proofs of its consistency, semantic completeness and decidability

Monadic predicate logic is the logic of one-place predicates. No two- or more place predicates appear.

*The language M*

The language M is like the language Q *except that it has no function symbols and its only predicate symbols are one-place predicate symbols.* Definitions of wff, etc., are the same as for Q, except that only individual constants and individual variables are terms, and clause 2 of the definition of wff is changed to

2′. If F is a predicate symbol and t is a term, then Ft is a wff and an atomic wff.

*Semantics of M*

As for Q.

*The system $QS^M$*

The axioms of $QS^M$ are those wffs *of M* that are axioms by any of the schemata QS 1–7. $QS^M$ has Modus Ponens for $\supset$ as its sole rule of inference.

50.1 *Every theorem of $QS^M$ is logically valid*

*Proof.* Every theorem of $QS^M$ is a theorem of QS, and every theorem of QS is logically valid.

50.2 *$QS^M$ is consistent*

*Proof.* From 50.1, by an already familiar argument.

50.3 (*The semantic completeness theorem for $QS^M$*) *Every logically valid wff of M is a theorem of $QS^M$*

*Proof.* By the semantic completeness of QS, every logically valid wff of M is a theorem of QS. So in order to prove the semantic completeness of $QS^M$, it is enough to show that for any proof in QS of a wff A of M there is a proof in $QS^M$ of A.

Let A be a wff of M that is logically valid. Then there is a proof in QS of A. Let $A_1, \ldots, A_m$ be a proof in QS of A (so $A_m$ is A). Wherever in this proof an $n$-place predicate symbol occurs followed by $n$ terms, with $n > 1$, replace it and its terms by the propositional symbol $p'$. The result will be a proof in $QS^M$ of A. For if $A_i'$ is the result of replacing in $A_i$ any polyadic predicate symbols and their associated terms by $p'$, then if $A_i$ is an axiom of QS $A_i'$ will be an axiom of $QS^M$: and if $A_k$ is an immediate consequence in QS of $A_j$ and $A_j \supset A_k$, then $A_k'$ is an immediate consequence in $QS^M$ of $A_j'$ and $(A_j \supset A_k)'$, because $(A_j \supset A_k)'$ is $(A_j' \supset A_k')$. So any logically valid wff of M has a proof in $QS^M$ and is therefore a theorem of $QS^M$.

50.4 *$QS^M$ is not negation-complete*

*Proof.* As for 46.4, replacing 'QS' by '$QS^M$'.

50.5 *$QS^M$ is not syntactically complete*

*Proof.* As for 46.5, substituting '$QS^M$' for 'QS' and 'M' for 'Q' throughout.

*The decidability of $QS^M$*

*Outline of the proof*

Let A be an arbitrary wff of M.

By 40.8 A is logically valid iff $A^c$ (an arbitrary closure of A) is. So to test the logical validity of A it is enough to test the logical validity of $A^c$.

We shall show – 50.8 – that a closed wff of M with $k$ distinct predicate symbols is logically valid iff it is $2^k$-valid.

By 40.22 there is an effective method for determining the $k$-validity of any formula of Q, and hence of any formula of M.

It follows that there is an effective method for determining the logical validity of any formula of M.

By 50.1 and 50.3 a formula of M is logically valid iff it is a theorem of $QS^M$.

So $QS^M$ is decidable.

The heart of the proof is Theorem 50.8, in which the decision problem for infinite domains is reduced to one for finite domains. 50.6 and 50.7 are lemmas for 50.8.

50.6 *If a closed wff of M is false for an interpretation I with a domain of k members, then it is false for some interpretation I′ with a domain of k + 1 members*

[50.6 in fact holds for arbitrary wffs of Q, but we prove it only for closed wffs of M.]

*Proof.* Let A be a closed wff of M with $n$ distinct predicate symbols that is false for some interpretation I with a domain of $k$ members.

Let the distinct predicate symbols in A be $F_1, \ldots, F_n$, and let I assign to them the properties $P_1, \ldots, P_n$. Let D be the domain of I.

We shall show that there is an interpretation I′, with a domain D′ of $k + 1$ members, for which A is false.

I′ is to be constructed from I as follows:

1. The domain of I′ is to be the domain of I with one extra object, b.
2. If d′ is a member of D′ and d′ ≠ b, then the member of D corresponding to d′ is to be d′ itself. If d′ = b, then the member of D corresponding to d′ is to be some (arbitrarily chosen) member of D.

3. If I assigns a member d of D to a constant c, then I′ also assigns d to c.
4. To the predicate symbols $F_1, \ldots, F_n$ I′ assigns the properties $P_1', \ldots, P_n'$ defined as follows:
   $P_1'$ belongs to a member d′ of D′ iff $P_1$ belongs to the member of D corresponding to d′. $P_2', \ldots, P_n'$ are defined similarly, but in terms of the properties $P_2, \ldots, P_n$.
5. I′ makes the same assignment of truth values to the propositional symbols of M as I does.

Next we show that the closed wff A is true for I iff it is true for I′. The proof is by induction on the number, *n*, of connectives and quantifiers in A.

*Basis*: $n = 0$

2 cases:

1. A is a propositional symbol. No trouble.
2. A is Fc where F is a monadic predicate symbol and c is a constant.
   1. Suppose A is true for I. Then the member d of D assigned by I to c has the property P. I′ also assigns d to c. By clause 4 above d has the property P′. So Fc, i.e. A, is true for I′. So if A is true for I it is true for I′.
   2. Suppose A is not true for I. Then the member d of D assigned by I to c does not have the property P. I′ also assigns d to c. By clause 4 d does not have the property P′. So Fc, i.e. A, is not true for I′. So if A is true for I′ it is true for I.

*Induction Step*

3 cases:

1. A is $\sim$B.
   1. Suppose A is true for I. Then B is not true for I. So by the induction hypothesis B is not true for I′. So (since B is closed) A is true for I′.
   2. Suppose A is true for I′. Then . . . (the rest is obvious)
2. A is $B \supset C$. (Since A is closed, B and C are.)
   1. Suppose A is true for I. Then since B and C are closed, either B is false for I or C is true for I. Then . . .
   2. . . .

3. A is $\Lambda$vB. (B may be open.)
   1. Suppose A is true for I. Then B is. Then by the induction hypothesis B is true for I′. Then A is.
   2. Suppose A is true for I′. Then B is. . . .

So we have: The closed wff A of M is true for I iff it is true for I′. So since A is *ex hypothesi* false for I, it is false for I′. But I was an arbitrary interpretation with a domain having an arbitrary number, *k*, of members ($k > 0$, of course). So for any interpretation with a domain of an arbitrary number of members in which an arbitrary closed wff of M is false, there is some interpretation with one more member in its domain for which the wff is false. Q.E.D.

The central idea of the proof of the following theorem is this: If you have *k* distinct monadic properties, and you consider only objects such that each of the properties either belongs or does not belong to each of the objects, then, no matter how many objects there are in the domain under consideration, the greatest number of logically possible different classes into which *k* monadic properties can in principle sort things in such a way that each thing is in one and only one class is exactly $2^k$. (If there are fewer than $2^k$ objects then at least one of the classes will be empty, but we are considering the number of possible pigeon-holes and not the number of pigeon-holes with pigeons in them).

For example: Suppose we have three monadic properties $P_1$, $P_2$ and $P_3$. Then the greatest possible number of classes into which these three properties can in principle sort things (in such a way that each thing is in one and only one class) is $2^3 = 8$. The classes are (writing 'are $P_1$' for 'have the property $P_1$' and 'are $\overline{P_1}$' for 'do not have the property $P_1$') the classes of:

1. Things that are $P_1$ and $P_2$ and $P_3$.
2. Things that are $\overline{P_1}$ and $P_2$ and $P_3$.
3. Things that are $P_1$ and $\overline{P_2}$ and $P_3$.
4. Things that are $\overline{P_1}$ and $\overline{P_2}$ and $P_3$.
5. Things that are $P_1$ and $P_2$ and $\overline{P_3}$.
6. Things that are $\overline{P_1}$ and $P_2$ and $\overline{P_3}$.
7. Things that are $P_1$ and $\overline{P_2}$ and $\overline{P_3}$.
8. Things that are $\overline{P_1}$ and $\overline{P_2}$ and $\overline{P_3}$.

No matter how many objects there are, if you have only those three monadic properties to play with, the greatest number of logically possible classes into which things can be sorted by their means (in such a way that each thing is in one and only one class) is $2^3 = 8$. And in general the number of ways in which $k$ monadic properties can in principle sort things is $2^k$.

So, to put it informally, if you have checked that a formula with $k$ distinct monadic predicate symbols (and no other predicate symbols) is true for anything in any of the appropriate $2^k$ classes, then there aren't any further different relevant possibilities that you need to consider (the number of pigeon-holes does not increase with the number of objects to be put in them).

A logically valid formula of M can be seen as a truth about *arbitrary abstract classes* of things; and in showing that monadic predicate logic is decidable we are at the same time showing that the simple calculus of classes is decidable.

50.7 *If a closed wff of M with k distinct predicate symbols is false for some interpretation I, then it is false for some interpretation I′ with a domain of at most $2^k$ members*

*Proof.* Let C be a closed wff of M, with $k$ distinct predicate symbols, that is false for some interpretation I.

We shall show that there is an interpretation I′ with a domain of at most $2^k$ members for which C is false.

Let the distinct predicate symbols in C be $F_1, \ldots, F_k$, and let I assign to them the properties $P_1, \ldots, P_k$ respectively.

We distribute the members of the domain D of I into non-overlapping *non-empty* classes defined by means of the properties $P_1, \ldots, P_k$, as illustrated in the example in the informal discussion preceding this proof. As the discussion showed, there can be at most $2^k$ such classes. (If there are fewer than $2^k$, this is because some combination of properties does not belong to any member of D.)

I′ takes these *classes* as the members of its domain D′.

Example: Let C be the closed wff $\Lambda x(Fx \supset Gx)$. Let I be an interpretation whose domain is the set of all animals and which assigns the property of *being a mammal* to $F$ and the property of *being a cow* to $G$. (So C is false for I.) Then I′ takes as members of its domain D′ the following classes:

1. The class of all animals that are both mammals and cows, i.e. the class of cows.
2. The class of all animals that are mammals but not cows.
3. The class of all animals that are neither mammals nor cows.

The class of all animals that are not mammals but are cows is not a member of the domain of I′, since this class is empty.

Let $P_1'$ be the property defined as follows: The property $P_1'$ belongs to a member d′ of D′ iff the property $P_1$ belongs to all the members of d′ (remember d′ is a class). $P_2'$, . . ., $P_k'$ are defined similarly, but in terms of the properties $P_2$, . . ., $P_k$ respectively. (Example below)

I′ assigns $P_1'$, . . ., $P_k'$ to the predicate symbols $F_1$, . . ., $F_k$ respectively, and arbitrary properties, defined for D′, to the other predicate symbols of M.

Example: Let C and I be as in the previous example. Then I′ assigns to *F* the property *having only mammals as members* and to *G* the property *having only cows as members.*

If I assigns a member d of D to a constant c, then I′ assigns to c the class in D′ to which d belongs.

Example: Let I be as before and let C be $Gc_1$, and let I assign to the constant $c_1$ a particular cow, say Bluebell. Then I′ assigns to $c_1$ the class in D′ to which Bluebell belongs, i.e. the first class in the list given above, i.e. the class of all animals that are both mammals and cows, i.e. the class of cows.

Finally I′ makes the same assignment of truth values to propositional symbols as I does.

We shall now show that in every case C is true for I iff it is true for I′. The proof is by induction on the number, *n*, of connectives and quantifiers in C.

*Basis*: $n = 0$

2 cases:

1. C is a propositional symbol. No trouble.
2. C is $F_i$c, where c is a constant and $1 \leqslant i \leqslant k$, Let I assign d in D to the constant c. Then I′ assigns to c the class, say d′, in D′ to which d belongs. Notice that it follows from our definition

of d′ that if d has the property $P_i$ $(1 \leqslant i \leqslant k)$, then every member of d′ has the property $P_i$, and that if d does not have the property $P_i$ $(1 \leqslant i \leqslant k)$, then no member of d′ has the property $P_i$.

(*a*) Suppose C is true for I. Then d has the property $P_i$ $(1 \leqslant i \leqslant k)$, and so, by the way d′ has been defined (see above), every member of the class d′ has the property $P_i$. So d′ has the property $P_i'$. So C is true for I′. So: If C is true for I it is true for I′.

(*b*) Suppose C is not true for I. Then d does not have the property $P_i$ $(1 \leqslant i \leqslant k)$. Then, by the way d′ has been defined (see above), no member of d′ has the property $P_i$. So d′ does not have the property $P_i'$. So C is not true for I′. So: If C is true for I′ it is true for I.

*Induction Step*

As in the Induction Step in the proof by induction in the proof of 50.6.

So C is false for I′. So, for any interpretation with a domain of arbitrary size for which an arbitrary closed wff of M with $k$ distinct predicate symbols is false, there is an interpretation with a domain of at most $2^k$ members for which the wff is false.

Q.E.D.

50.8 *A closed wff of M with k distinct predicate symbols is logically valid iff it is $2^k$-valid*

*Proof.* Let C be a closed wff of M with $k$ distinct predicate symbols.

1. Obviously if C is logically valid C is $2^k$-valid.

2. Suppose C is not logically valid. Then it is false for some interpretation I. Then by 50.7 it is false for some interpretation I′ with a domain of at most $2^k$ members. Then by 50.6, applied as many times as may be necessary, it is false for some interpretation with a domain of exactly $2^k$ members: i.e. it is not $2^k$-valid. So if C is $2^k$-valid it is logically valid.

The decidability of $QS^M$ follows, as in the *Outline of the proof* on p. 210. So:

50.9 *$QS^M$ is decidable*

PART FOUR

# First Order Predicate Logic: Undecidability

## 51 Some results about undecidability[1]

In this section we prove various results about decidability, using only the informal notion of effective method explained in §7.

Our first concern is to establish an upper limit to the number of effective methods there are.

It would be generally agreed that any effective method (in logic and mathematics) must be capable of being unambiguously specified by means of a finite string of words and/or symbols. If it could not be finitely specified, then it could not even in principle be written out as a program for an actual or possible computer, and so it would not be a method of computing.

We shall make a stronger assumption, viz. that some existing language (such as English or French or German), together with the (interpreted) symbols of existing logic and mathematics, and supplemented if necessary by a finite set of new words and/or new (interpreted) symbols, is adequate for the unambiguous specification of all effective methods.

In *partial* justification of this assumption (which is a primitive form of Church's Thesis: cf. §52 below):

1. We simply have no idea of an effective method (in logic and mathematics) that could not in principle be specified by *any* finite extension of *any* existing natural language, supplemented by logical and mathematical symbolism.
2. So far as logic and mathematics are concerned, whatever can be said in some existing natural language or other can be said in some *one* existing natural language, such as English or French or German.

English is as good a natural language as any to take here, so we shall state the assumption in the following form:

51.1 *Assumption. English, supplemented if necessary by a finite set of (interpreted) symbols and by a finite set of new words or words borrowed from other languages, is adequate for the unambiguous specification of all effective methods*

[1] This section borrows a good deal from Shepherdson (1967, esp. p. 3 and fn.).

By Assumption 51.1, for each distinct effective method there is at least one distinct (finite) string of words in English, or supplemented English, that unambiguously specifies it. Such a string is composed of letters and symbols from a finite alphabet, and there are only countably many finite strings of symbols from a finite alphabet (cf. the proof of 14.1). So:

51.2 *There are only countably many effective methods*

It might be objected that a string is in principle capable of being interpreted in infinitely many different ways, and so, though there are only countably many different strings, they could in principle be used to say uncountably many different things. However, this objection is taken care of by our requirement (Assumption 51.1) that for each effective method there is an *unambiguous* specification. A string that could be interpreted either as a description of one effective method or as a description of quite a different one is not an unambiguous specification of either method. So, when we are calculating how many effective methods there are, Assumption 51.1 allows us to ignore strings that have more than one interpretation.

51.3 *There are sets of natural numbers that are not effectively enumerable*

*Proof.* There are uncountably many subsets of the set of natural numbers (§11). By 51.2 there are only countably many effective methods. Any effective method for enumerating a set of natural numbers enumerates just one set of natural numbers. So there are some sets of natural numbers for whose enumeration there is no effective method.

This is a remarkable result. The sets that cannot be effectively enumerated are *countable* sets, not uncountable ones; and they are sets of natural numbers, so for each set one can speak of 'the smallest number in the set', 'the next greatest number in the set', 'the *n*th smallest number in the set', and so on. Yet, despite that, there is no effective method for enumerating their members.

51.4 *There are undecidable sets of natural numbers*

*Proof.* There are uncountably many subsets of the set of natural numbers, but only countably many effective methods. So there are sets of natural numbers for which there is no effective method for telling whether or not an arbitrary natural number is a member.

*Definition.* The *complement* of a set S is the set of all things that are not members of the set S. The *complement of a set S with respect to, or relative to, a set T* is the set of all members of T that are not members of S.

If S is a set of natural numbers, by the *relative complement* of S we shall mean the set of all natural numbers that are not members of S [in symbols: $\overline{S}$].

51.5 *A set of natural numbers is decidable iff both it and its relative complement are effectively enumerable* [Post, 1944]

*Proof.* Let S be a set of natural numbers.

1. Suppose S is decidable.

(*a*) If S is empty, then S is effectively enumerable (by convention) and $\langle 0, 1, 2, 3, 4, \ldots \rangle$ is an effective enumeration of $\overline{S}$. Similarly if $\overline{S}$ is empty.

(*b*) If S is not empty, then $\langle a_1, a_2, a_3, \ldots \rangle$ is an effective enumeration of S, where $a_1$ is the smallest natural number in S, $a_2$ the next greatest in S, and so on; and $\langle b_1, b_2, b_3, \ldots \rangle$ is an effective enumeration of $\overline{S}$, where $b_1$ is the smallest natural number not in S, $b_2$ the next greatest not in S, and so on. (Either of these sets may be finite.) These enumerations are effective because (by our assumption) S is decidable. [Informally: Test 0 to see if it is in S. If it is, set it down as such. If it is not, set it down as a member of $\overline{S}$. (Since S is decidable, there is an effective method for telling whether or not 0 is a member of S.) Repeat the procedure for 1. And so on. The resulting sequences are effective enumerations of S and $\overline{S}$.]

So if S is decidable, S and $\overline{S}$ are effectively enumerable.

2. Suppose S and $\overline{S}$ are effectively enumerable.

(*a*) If S is finite, S is decidable (§8). If $\overline{S}$ is finite, $\overline{S}$ is decidable. Any effective method for determining membership of $\overline{S}$ can be converted into an effective method for determining membership of S by simply taking the opposite answer to the one given by the test for membership of $\overline{S}$. So if $\overline{S}$ is finite, S is decidable.

(*b*) Suppose S and $\overline{S}$ are both infinite. Let $\langle a_1, a_2, a_3, \ldots \rangle$ and $\langle b_1, b_2, b_3, \ldots \rangle$ be effective enumerations of S and $\overline{S}$ respectively. Then any natural number *n* occurs in one or other, but not both, of those enumerations. To test *n* for membership of S, consider in order the terms $a_1$, $b_1$, $a_2$, $b_2$, $a_3$, $b_3$, . . ., keeping track of whether they are in S or in $\bar{S}$. Sooner or later *n* is bound to turn up in this list, and then it will be known whether it is a member of S or of $\bar{S}$.

So if S and $\bar{S}$ are both effectively enumerable, S is decidable.

*Definition.* A function of *n* arguments from natural numbers to natural numbers is *total* iff its domain is the set of *all* ordered *n*-tuples of natural numbers (if $n = 1$, its domain is the set of *all* natural numbers).

*Definition.* A function of *n* arguments from natural numbers to natural numbers is *computable* iff it is total and there is an effective method for computing the value of the function for each member of its domain.

Examples: The sum function for natural numbers is total, and there is an effective method for computing the sum of any two natural numbers, so the sum function for natural numbers is a computable function. The product function for natural numbers is another computable function.

### 51.6 *There are only denumerably many computable functions from natural numbers to natural numbers*

*Proof.* It is obvious that there are at least denumerably many: consider, e.g., the sequence of functions defined by the rules $f_0(x) = x + 0$, $f_1(x) = x + 1$, $f_2(x) = x + 2$, . . . Since there are only countably many effective methods, there are at most countably many computable functions from natural numbers to natural numbers. So there are just denumerably many of them.

### 51.7 *There are non-computable total functions from natural numbers to natural numbers*

*Proof.* To each distinct non-empty subset of the set of natural numbers there corresponds a distinct total function from natural numbers to natural numbers, viz. the function whose value for the argument 0 is the smallest number in the subset in question, whose value for the argument 1 is the next greatest number in the set if there is one and otherwise is the same value as for the argument 0; and so on. There are uncountably many non-empty subsets of the set of natural numbers (omitting the empty set still leaves uncountably many subsets: cf. 13.6). So there are uncountably many of those functions. But there are only countably many effective methods. So there are non-computable total functions from natural numbers to natural numbers.

51.8 *The total function g from natural numbers to natural numbers, defined below, is not computable*

*Proof.* By 51.6 the set of all computable functions from natural numbers to natural numbers is denumerable. Clearly the set of all computable functions of one argument from natural numbers to natural numbers is not finite, so it is also denumerable. Let $\langle f_0, f_1, f_2, \ldots \rangle$ be an enumeration of it (this enumeration is not effective, as we shall prove in 51.9). Let $g$ be the total function from natural numbers to natural numbers defined by the rule

$$\begin{cases} g(n) = 1 \text{ if } f_n(n) = 0 \\ g(n) = 0 \text{ otherwise.} \end{cases}$$

Then $g$ is not computable. For if it were, it would be one of the $f_i$'s, say $f_k$. Then we would have

$$g(k) = f_k(k) \quad \ldots \quad \ldots \quad \ldots \quad (1)$$

But by the definition of $g$ if $f_k(k) = 0$ then $g(k) = 1$ and if $f_k(k) \neq 0$ then $g(k) = 0$. So $g(k) \neq f_k(k)$, which contradicts (1). So $g$ is not computable.

51.9 *The set of computable functions of one argument from natural numbers to natural numbers is not effectively enumerable*

*Proof.* Suppose it were. Then there would be an effective method for finding, given any natural numbers $j$ and $k$, the $j$th function in the enumeration and the value of the $j$th function for the argument $k$. So there would be an effective method for computing the value of $f_n(n)$ for each natural number $n$. So there would be an effective method for computing the value of $g(n)$ for each $n$, where $g$ is the function defined in 51.8. So $g$ would be computable. But $g$ is not computable, as we proved. So the set of computable functions of one argument from natural numbers to natural numbers is not effectively enumerable.

*Definitions*

We shall say that any formal system satisfying the following conditions is a *formal system of arithmetic*:

1. On its intended interpretation at least some of its theorems express truths of pure number theory.
2. For each natural number $n$ there is in the system an expression in some standard form that on the intended interpretation denotes the number $n$. We shall call this expression the *numeral* for $n$, and we denote it, in the metalanguage, by $\bar{n}$. E.g. $\bar{3}$ (the numeral for the number 3) might be '3' or '111' or 'SSS0' or '0‴'.
3. The formulas of the system are finite strings of symbols from a finite alphabet.

We shall say that a formal system of arithmetic is *respectable* if the following conditions are satisfied:

1. The system is consistent.
2. Every decidable set of natural numbers is represented in it, in a sense of 'represented' to be explained immediately below.
3. A formula with free variables is a theorem iff some closure of it is.

We shall say that a set X of natural numbers is *represented* in a formal system S iff there is a formula A(v) of S with just one free variable v, such that for each natural number $n$

$$\vdash_S A(\bar{n}) \quad \text{iff} \quad n \in X$$

where A($\bar{n}$) is the result of substituting the numeral $\bar{n}$ for all free occurrences of the variable v in A(v). [Or, using a notation we used earlier: . . . iff there is a formula A of S with just one free variable v such that for each natural number $n$

$$\vdash_S A\bar{n}/v \quad \text{iff} \quad n \in X]$$

['$n \in X$' means '$n$ is a member of the set X'.]

51.10 *Any respectable formal system of arithmetic is undecidable*

*Proof.* Let S be a formal system of arithmetic. Each formula A of S determines a function $f$ as follows:

$$\begin{cases} f(n) = 0 & \text{if the formula obtained by replacing all free occurrences of variables in A by the numeral } \bar{n} \text{ is a theorem of S} \\ f(n) = 1 & \text{otherwise} \end{cases}$$

Condition 3 of the definition of *formal system of arithmetic* ensures that there is an effective enumeration of the formulas of S: cf. the proof of 32.11. Let $\langle A_0, A_1, A_2, \ldots \rangle$ be an effective enumeration of the formulas of S, and let $\langle f_0, f_1, f_2, \ldots \rangle$ be the corresponding enumeration of the functions determined by the formulas $A_0, A_1, A_2, \ldots$ Let $h$ be the function defined by the rule

$$\begin{cases} h(n) = 1 \text{ if } f_n(n) = 0 \\ h(n) = 0 \text{ otherwise} \end{cases}$$

Then $h$ is not in the enumeration $\langle f_0, f_1, f_2, \ldots \rangle$ For if it were, then for some natural number $k$ we should have $f_k(k) = 1$ iff $f_k(k) = 0$.

We define a set H of natural numbers as follows:

$$n \in \mathrm{H} \quad \text{iff} \quad h(n) = 0$$

Now *if the system S is decidable then the set H is decidable.* For suppose that S is decidable; then each of the $f_i$'s is computable, and the value of $f_j(k)$ is computable for any pair of natural numbers $j$ and $k$; so $h$ is computable; so H is decidable.

But *the set H is not represented in S.* For suppose it were. Then by the definition of representation there would be a formula A(v) of S such that

$$\vdash_S A(\bar{n}) \quad \text{iff} \quad n \in \mathrm{H}$$

But we have, by the definition of H,

$$n \in \mathrm{H} \quad \text{iff} \quad h(n) = 0$$

So we should have

$$\vdash_S A(\bar{n}) \quad \text{iff} \quad h(n) = 0$$

But then the formula A(v) would determine the function $h$, and so $h$ would be in the enumeration of the functions corresponding to the enumeration of the formulas of S. But we showed that $h$ cannot be in that enumeration. So H cannot be represented in S.

So *if S is decidable then there is a decidable set of natural numbers not represented in S* (viz. the set H): i.e. if S is decidable then S is not respectable. So if S is respectable S is undecidable.

Q.E.D.

*Gödel numbering*

A Gödel numbering is an assignment of numbers, called 'Gödel numbers', to symbols, (finite) sequences of symbols, and (finite) sequences of (finite) sequences of symbols in such a way that (1) there is a 1–1 correspondence between the set of Gödel numbers on the one hand and the set of symbols, sequences of symbols, and sequences of sequences of symbols on the other, and (2) there is an effective method, given a symbol, sequence of symbols, or sequence of sequences of symbols, for finding its Gödel number, and an effective method, given a number, for finding whether it is a Gödel number and if so which symbol, sequence of symbols, or sequence of sequences of symbols it is the Gödel number of.

In standard formal systems formulas will be certain finite sequences of symbols, and proofs will be certain finite sequences of finite sequences of symbols. Given a Gödel numbering for such a system, each distinct formula and each distinct proof will have its own distinct and unique Gödel number.

There are many different ways of assigning numbers so as to achieve a Gödel numbering. Here is one.

To each distinct symbol we assign as its Gödel number a distinct odd positive integer other than 1. For example, suppose the symbols are just 'A', 'B' and 'C': to 'A' we assign the number 3, to 'B' the number 5, to 'C' the number 7.

To an *n*-termed sequence of symbols we assign as its Gödel number

$$2^{s_1} \times 3^{s_2} \times 5^{s_3} \times \ldots \times p_n{}^{s_n}$$

where $s_1$ is the Gödel number of the first symbol in the sequence, $s_2$ is the Gödel number of the second symbol in the sequence, and so on, 2, 3 and 5 are the first three prime numbers, and $p_n$ is the *n*th prime number. For example, where 'A', 'B' and 'C' have the Gödel numbers 3, 5 and 7 respectively, the Gödel number of the sequence ⟨'C', 'A', 'B'⟩ will be

$$2^7 \times 3^3 \times 5^5$$

and the Gödel number of the sequence ⟨'B', 'A', 'C', 'C', 'A'⟩ will be

$$2^5 \times 3^3 \times 5^7 \times 7^7 \times 11^3.$$

Notice that symbols will have odd Gödel numbers, sequences of symbols even Gödel numbers. Formulas count as sequences, so a formula consisting of just one symbol will have a different Gödel number from the symbol of which it consists.

Finally, we assign Gödel numbers to sequences of sequences

of symbols. Suppose that $e_1, e_2, \ldots, e_m$ are sequences of symbols and that the Gödel numbers of these sequences are $g_1, g_2, \ldots, g_m$ respectively. Then the Gödel number of the sequence $\langle e_1, e_2, \ldots, e_m \rangle$ is to be

$$2^{g_1} \times 3^{g_2} \times \ldots \times p_m{}^{g_m}$$

where $p_m$ is the *m*th prime number.

The Gödel number of a sequence of sequences of symbols will be different from the Gödel number of any sequence of symbols, for the exponent of 2 in its prime factorisation will be an even number, while the exponent of 2 in the prime factorisation of the Gödel number of a sequence of symbols will be odd.

51.11 *If the set of formulas of a formal system S is decidable, and there is an effective method for telling whether or not something is a proof in S, then the set of theorems of S is effectively enumerable*

*Proof.* Let S be a formal system satisfying all the hypotheses of the theorem. Since the set of formulas of S is decidable, the formulas must be finite sequences of symbols, and the set of all symbols actually occurring in the formulas must be decidable. The symbols, finite sequences of symbols, and finite sequences of finite sequences of symbols of this decidable set admit of a Gödel numbering. The Gödel numbering and the decidability of the set of formulas of S together ensure that the set of all finite sequences of formulas of S is effectively enumerable. Since there is an effective method for telling whether or not something is a proof in S, a proof in S must be finitely long, and so it will be a finite sequence of formulas of S. Let $\langle S_1, S_2, S_3, \ldots \rangle$ be an effective enumeration of all finite sequences of formulas of S. Then we can get an effective enumeration of the theorems of S as follows: Test $S_1$ to see whether it is a proof in S. If it is, set down the theorem of which it is a proof as the first term in a sequence of theorems (to be constructed). Then test $S_2$. If it is a proof, set down the theorem of which it is a proof as the next term in the sequence of theorems (or as the first term, if $S_1$ was not a proof in S). And so on, in the obvious way. The resulting sequence is an effective enumeration of the theorems of S. (If S has no theorems, then the sequence is empty. But by convention an empty sequence counts as an effective enumeration of the empty set, and by convention the empty set is effectively enumerable.)

51.12 *Let S be a formal system in which any formula with free variables is a theorem iff an arbitrary closure of it is. If the set of formulas of S is decidable and there is an effective method for telling whether or not something is a proof in S, then if S is consistent and negation-complete, then S is decidable*

*Proof.* Let S be a formal system satisfying all the hypotheses of the theorem. Let A be an arbitrary formula of S which we want to test for theoremhood. Let $A^c$ be an arbitrary closure of A. By 51.11 the theorems of S are effectively enumerable. Since S is negation-complete, either $A^c$ or the negation of $A^c$ occurs in such an enumeration. If $A^c$ does, then A is a theorem. If the negation of $A^c$ does, then since S is consistent, $A^c$ is not a theorem, and so A is not a theorem.

*Note:* If S is absolutely inconsistent, but satisfies the other hypotheses, then S is decidable. For if S is absolutely inconsistent then the theorems of S coincide with the formulas of S, and by hypothesis the set of formulas of S is decidable.

51.13 (*The Generalised Gödel Theorem*) *Let S be any respectable formal system of arithmetic. If S has a decidable set of formulas and there is an effective method for telling whether or not something is a proof in S, then S is incomplete* (*i.e. not negation-complete*)

*Proof.* Directly from 51.12, 51.10, and the definition of respectability.

If every closed wff of the interpreted system is a (true or false) proposition of arithmetic, then the system will be incomplete in a semantic sense: some *true* proposition of arithmetic will not be a theorem of the system.

*Comment.* It would be a mistake to draw the conclusion that 'any consistent axiomatization of the theory of the natural numbers must always fail to capture as theorems all the truths about the natural numbers' (this is a quotation from a book on the philosophy of mathematics). For the following is a consistent axiomatisation of number theory that captures all the truths of pure number theory:

System C

Axioms: All truths of pure number theory.
Rules of inference: None.

(Of course C is not a *formal* system.) The point is rather that no consistent axiomatic system *with a decidable set of axioms* can capture all the truths of pure number theory.

Again, it would be a mistake to conclude that there is *some particular true proposition* of pure number theory that cannot ever be a theorem of a consistent system. Let T be an arbitrary true proposition of pure number theory. Then any system that has T as its sole axiom, and either no rules of inference or truth-preserving rules of inference, is consistent and has T as a theorem.

For brevity let us call a formal system that differs from QS [$QS^=$] at most in having a finite set of axioms that are formulas of Q but not axioms of QS [$QS^=$] a *finite extension* of QS [$QS^=$].

51.14 *If some finite extension of QS* [$QS^=$] *is undecidable, then so is QS* [$QS^=$]

*Proof.* Let S be a finite extension of QS. Let the axioms of S that are not axioms of QS be $A_1, \ldots, A_n$. Then $\vdash_S B$ for an arbitrary formula B iff $(A_1 \wedge \ldots \wedge A_n) \vdash_{QS} B$, where $(A \wedge B)$ is an abbreviation for $\sim(A \supset \sim B)$, and brackets are added as required. But by the Deduction Theorem and its converse [43.1, 43.2] $A \vdash_{QS} B$ iff $\vdash_{QS} A \supset B$. So $(A_1 \wedge \ldots \wedge A_n) \vdash_{QS} B$ iff $\vdash_{QS} (A_1 \wedge \ldots \wedge A_n) \supset B$. So $\vdash_S B$ iff $\vdash_{QS} (A_1 \wedge \ldots \wedge A_n) \supset B$. So if QS were decidable, S would be too. So if S is undecidable, so is QS.

Writing '$QS^=$' for 'QS' throughout this proof we get the result: If S is a formal system that differs from $QS^=$ at most in having a finite set of axioms that are formulas of Q but not axioms of $QS^=$, then if S is undecidable so is $QS^=$.

51.15 *If there is a finite extension of QS* [$QS^=$] *that satisfies the conditions*

1. *All its extra axioms are closed wffs;*
2. *It is a formal system of arithmetic;*
3. *It is consistent;*
4. *Every decidable set of natural numbers is represented in it:*

*then QS* [$QS^=$] *is undecidable*

*Proof.* Suppose S is such a system. Then condition 1 ensures that S is a first order theory, and so (since by 45.5 for any first order theory K $\vdash_K$ A iff $\vdash_K$ $A^c$) S satisfies the third condition for respectability, viz. that a formula with free variables is a theorem iff some closure of it is. Conditions 2–4 ensure that S satisfies the other conditions for being a respectable formal system of arithmetic. By 51.10 any such system is undecidable. S is *ex hypothesi* a finite extension of QS [$QS^=$]. So by 51.14 QS [$QS^=$] is undecidable [remember that this is all under the supposition that there is such a system S]. So *if* there is such a system, QS [$QS^=$] is undecidable.

This is as far as we can get towards the undecidability of first order predicate logic without hard work and an appeal to Church's Thesis (explained in §52).

What we are going to do is present a formal system of arithmetic that is a finite extension of $QS^=$, prove that every *recursive* set of natural numbers is represented in the system (this is hard work), and then appeal to Church's Thesis in the form

*Every decidable set of natural numbers is recursive.*

## 52 Church's Thesis (1935). Church's Theorem (1936)

Briefly, Church's Thesis is the proposal that we should identify the informal, intuitive notion of effectiveness (in logic and mathematics) with the precisely defined mathematical notion of recursiveness (explained in §53). This is not something that admits of rigorous formal proof. Rather, it is a challenge to produce something that is effective, in the intuitive sense, but is not recursive. No one has yet done this.

Kurt Gödel in his 1931 paper gave a precise definition of a class of computable number-theoretic functions that he called 'recursive functions' and that are now called, after Kleene, 'primitive recursive functions'. In lectures in 1934, following a suggestion of Herbrand's, he defined a wider class of computable functions, now known as 'general recursive', or simply 'recursive', functions. Pretty well simultaneously [1932–5] Alonzo Church and S. C. Kleene were working on a precisely defined class of computable functions that they called 'λ-definable functions' [lambda-definable functions]. It seemed to them that this class might well embrace all functions that can be regarded as effectively computable in the informal sense, and they proved that the class of λ-definable functions is the same as the class of general recursive functions. Accordingly Church proposed, in 1935, that the informal notion of an effectively calculable function of positive integers should be identified with the precisely defined mathematical notion of a (general) recursive function (*Bulletin of the American Mathematical Society*, May 1935, [vol. 41] p. 333; received for publication 22 March 1935). In 1936 Alan M. Turing gave a quite different kind of definition of a class of computable functions, in terms of what have since come to be known as Turing machines; and he proved that his definition of computability was equivalent to Church's, in the sense that a function is Turing computable iff it is λ-definable (or, equivalently, general recursive). Emil Post, in the same year and independently of Turing, published a brief analysis of computability that was fundamentally the same as Turing's. In a paper published in 1943 Post gave two further analyses, one in terms of Post combinatorial systems, and one in terms of Post normal systems: some of the ideas in this paper came from his own unpublished work of 1920–2. Other analyses followed: e.g. Andrei Markov's theory of algorithms (1951). All these different definitions and analyses, from Gödel's 1934 one on, have turned out to be equivalent, in the sense that they all define exactly the same class of number-theoretic functions. In view of these equivalences, and because Church's Thesis in one direction is regarded as obvious ('Every recursive function is effectively computable'), the Thesis will be found stated in various different ways.

In the paper of which the note in the *Bulletin* was an abstract Church proved that, under the assumption of the Thesis, there is no effective method for the solution of a certain class of problems of elementary number theory [Church, 1935*b*]. This was the first ever result of this kind. In the following year, 1936, Church showed, again under the assumption of the Thesis, that Hilbert and Ackermann's system of first order predicate logic is undecidable [Church, 1936]. This last result is known as *Church's Theorem*, and it is quite different from his Thesis. The Theorem is capable of rigorous proof, given the Thesis. Or it can be stated and proved without any appeal to the Thesis in either of the following forms:

Either:

The decision problem for Hilbert and Ackermann's system of first order predicate logic is *recursively* unsolvable.

Or in a more general form:

The set of logically valid formulas of any language adequate for first order predicate logic is not *recursive*.

(This last formulation makes use of the fact that the formulas of such a language can be numbered.)

Our plan, of showing that every recursive set of natural numbers is represented in a certain formal system, requires that we should next give a brief account of recursive function theory.

## 53 Recursive functions.[1] Recursive sets

We now define a class of functions whose arguments and values are natural numbers. The intention is that the class should contain all and only such functions as are, in an intuitive sense, computable. We start with an initial stock of functions that are clearly computable, specify some operations that when applied to computable functions produce computable functions, and then define the class as the class of all functions obtainable from the initial stock by applying those operations a finite number of times.

We are concerned exclusively with functions whose arguments and values are natural numbers, and this is to be understood throughout what follows.

Recursive functions were used in research on the foundations of mathematics by Dedekind [1887], Peano [1889], Skolem [1923], Hilbert [1925, 1927] and Ackermann [1927]. They got their name from Gödel's 1931 paper, where his definition of the class of (primitive) recursive functions refers to functions 'recursively defined' from others. A recursive definition of a function is in effect a definition by mathematical induction: very roughly, it consists of a set of equations, one of which

[1] This account of recursive functions leans heavily on Andrzej Grzegorczyk (1961, chap. I). Fuller accounts of recursive functions may be found in Péter (1951), Kleene (1952), Hermes (1961), Rogers (1967). Cf. also Davis (1958) and, on algorithms generally, Markov (1954).

gives the value of the function for the argument 0, while the others say what the value of the function is for the argument $k+1$ in terms of its value for the argument $k$. So the value for any given argument can be computed by starting with 0 and applying the recursion equations as often as may be necessary. However, it turns out that explicit mention of recursion is not essential to the definition of recursive functions, and we shall not mention it in our definition below. Also, since we are concerned with recursive functions for a particular purpose and not for their own sake, we shall not stop to define the subclass of recursive functions known as primitive recursive functions, or the superclass known as partial recursive functions.

*Initial functions*

1. The successor function, $f_1$, defined by the rule $f_1(x) = x + 1$
2. The sum function, $f_2$, defined by the rule $f_2(x, y) = x + y$
3. The product function, $f_3$, defined by the rule $f_3(x, y) = x . y$
4. The power function, $f_4$, defined by the rule $f_4(x, y) = x^y$ [1]
5. The arithmetic difference function, $f_5$, defined by the rule $f_5(x, y) = x \dot{-} y$
   [$x \dot{-} y$ is known as the *arithmetic difference* between $x$ and $y$. If $x > y$, $x \dot{-} y$ is the ordinary difference. If $x \leqslant y$, then $x \dot{-} y = 0$.]

It is to be understood that all these functions are *total* functions from natural numbers to natural numbers.

*Operations that when applied to computable functions produce computable functions*

1. *Composition*

Let $f$ and $g$ be functions. Then the function $h$ defined by the rule

$$h(\ldots, x, \ldots, y, \ldots) = f(\ldots, x, \ldots, g(\ldots, y, \ldots), \ldots)$$

is obtained by *composition* from the functions $f$ and $g$. It is clear that if $f$ and $g$ are computable, then $h$ will also be computable.

2. *The $\mu$-operation* [*the mu-operation*]

Let $f(x_1, \ldots, x_n, y)$ be a *computable* function of $n+1$ arguments *such that for each n-tuple of natural numbers* $\langle x_1, \ldots, x_n \rangle$ *there*

[1] To avoid formal complications later, we set $0^0$, nomally left undefined, as, by definition, equal to 1.

*is a natural number $y$ such that* $f(x_1, \ldots, x_n, y) = 0$. Then the function $g$ defined by the rule

$$g(x_1, \ldots, x_n) = \mu y\{f(x_1, \ldots, x_n, y) = 0\},$$

where the right-hand side means 'The least number $y$ such that $f(x_1, \ldots, x_n, y) = 0$', is obtained by the *μ-operation* from the function $f$.

It is clear that when the μ-operation is applied to a function $f$ of $n+1$ arguments satisfying the two conditions (1) that $f$ is computable and (2) that for each $n$-tuple of natural numbers there is a natural number $y$ such that $f(x_1, \ldots, x_n, y) = 0$, then the resulting function is also computable.

When we speak of 'functions obtainable by the μ-operation' we shall always mean functions obtainable by the μ-operation applied to functions *satisfying the two conditions.*

*Definition.* The class of *recursive functions* is the class of all functions obtainable from the initial functions by a finite number of applications of the μ-operation and/or composition. All these functions will be total functions from natural numbers to natural numbers.

[This class is identical with the class of, for example, Turing computable functions and with the class of functions computable by Markov algorithms.]

*Recursive sets*

We are still concerned exclusively with sets of natural numbers.

A set X of natural numbers is *recursive* iff there is a recursive function $f$ such that for each natural number $n$, $n \in X$ iff $f(n) = 0$.[1]

Since every recursive function is computable, it follows that if a set is recursive then there is an effective method for determining whether or not an alleged member of the set really is a member: i.e. every recursive set is decidable.

## 54 Representation, strong representation and definability of functions in a formal system

We shall say that a function $f$ (from natural numbers to natural numbers) is *represented* in a formal system S iff there is a formula of S, which we shall write as $A(v_1, \ldots, v_{n+1})$, with free

[1] '$n \in X$' means '$n$ is a member of the set X'.

variables $v_1, \ldots, v_{n+1}$, such that for each $n+1$-tuple of natural numbers $\langle k_1, \ldots, k_{n+1}\rangle$

$$f(k_1, \ldots, k_n) = k_{n+1} \text{ iff } \vdash_S A(\overline{k_1}, \ldots, \overline{k_{n+1}}),$$

where $A(\overline{k_1}, \ldots, \overline{k_{n+1}})$ is the result of substituting the numerals $\overline{k_1}, \ldots, \overline{k_{n+1}}$ for the variables $v_1, \ldots, v_{n+1}$ respectively in $A(v_1, \ldots, v_{n+1})$.

If every recursive function is represented in S, then every recursive set is too. For let X be an arbitrary recursive set. Then by definition there is a recursive function $f$ such that, for each natural number $n$, $f(n) = 0$ iff $n \in X$. Let S be a formal system in which every recursive function is represented. Then $f$ is represented in S; i.e. there is a formula $A(v_1, v_2)$ of S such that for each pair of natural numbers $m, n$

$$f(m) = n \text{ iff } \vdash_S A(\bar{m}, \bar{n})$$

and therefore such that

$$f(m) = 0 \text{ iff } \vdash_S A(\bar{m}, \bar{0})$$

[$\bar{0}$ is the numeral in S for 0]

So we have:

$$m \in X \text{ iff } f(m) = 0 \text{ iff } \vdash_S A(\bar{m}, \bar{0})$$

But $A(v, \bar{0})$ is a formula with just one free variable, v. So there is a formula $A(v, \bar{0})$ of S with just one free variable such that, for each natural number $m$, $m \in X$ iff $\vdash_S A(\bar{m}, \bar{0})$: i.e. X is represented in S.

*Strong representation*

We shall say that a function $f$ of $n$ arguments is *strongly represented* in a formal system S iff there is a formula $A(v_1, \ldots, v_{n+1})$ of S with free variables $v_1, \ldots, v_{n+1}$ such that for each $n+1$-tuple of natural numbers $\langle k_1, \ldots, k_{n+1}\rangle$

(i) if $f(k_1, \ldots, k_n) = k_{n+1}$, then $\vdash_S A(\overline{k_1}, \ldots, \overline{k_{n+1}})$

*and*

(ii) if $f(k_1, \ldots, k_n) \neq k_{n+1}$, then $\vdash_S \sim A(\overline{k_1}, \ldots, \overline{k_{n+1}})$

It follows from our definitions that if a function $f$ is strongly represented in a system S and S is consistent, then $f$ is represented in S.

*Definability of functions in a formal system*

A function $f$ of one argument is *definable* in a formal system S iff it is strongly represented in S by a formula $A(x, y)$ and condition (iii) below is also satisfied: i.e. iff

(i) For all pairs of natural numbers $m$, $n$ if $f(m) = n$ then $\vdash_S A(\bar{m}, \bar{n})$

(ii) For all pairs of natural numbers $m$, $n$ if $f(\mathrm{m}) \neq n$ then $\vdash_S \sim A(\bar{m}, \bar{n})$

(iii) $\vdash_S \Lambda x \Lambda y \Lambda z(A(x, y) \supset (A(x, z) \supset y = z))$

This definition can easily be generalised to functions of more than one argument.

Clearly if $f$ is definable in S and S is consistent, then $f$ is represented in S.

## 55 A formal system of arithmetic: the system H

The next step is to present a formal system in which (as we shall show) every recursive set of natural numbers is represented.

The system H is a first order theory whose proper axioms are the axioms by the axiom-schema QS= 2 (§47) and the 24 numbered axioms below.

The language of H is the language Q. But for readability we write

| | | |
|---|---|---|
| $x, y, z, w$ | for | $x', x'', x''', x''''$ |
| $0$ | for | $a'$ |
| $\mathrm{St}$ | for | $f^{*\prime}\mathrm{t}$ |
| $\mathrm{t}_1 + \mathrm{t}_2$ | for | $f^{**\prime}\mathrm{t}_1\mathrm{t}_2$ |
| $\mathrm{t}_1 . \mathrm{t}_2$ | for | $f^{**\prime\prime}\mathrm{t}_1\mathrm{t}_2$ |
| $\mathrm{Pt}_1\mathrm{t}_2$ | for | $f^{**\prime\prime\prime}\mathrm{t}_1\mathrm{t}_2$ |
| $\mathrm{t}_1 = \mathrm{t}_2$ | for | $F^{**\prime}\mathrm{t}_1\mathrm{t}_2$ |
| $\mathrm{t}_1 < \mathrm{t}_2$ | for | $F^{**\prime\prime}\mathrm{t}_1\mathrm{t}_2$ |
| $(A \wedge B)$ | for | $\sim(A \supset \sim B)$ |
| $(A \vee B)$ | for | $(\sim A \supset B)$ |

and we use square brackets to indicate some groupings. (When the axioms are written out in full these square brackets are unnecessary.)

The domain of the intended interpretation of H is the set of natural numbers.

On the intended interpretation $a', f^{*\prime}a', f^{*\prime}f^{*\prime}a', f^{*\prime}f^{*\prime}f^{*\prime}a'$, . . . are numerals, which in our abbreviated notation we write as

$$0, S0, SS0, SSS0, \ldots$$

On the intended interpretation $f^{*\prime}$ (S) expresses the successor function; the function symbols $f^{**\prime}$, $f^{**\prime\prime}$, $f^{**\prime\prime\prime}$ express the sum function (addition), the product function (multiplication), and the power function (exponentiation) respectively; and the predicate symbols $F^{**\prime}$ and $F^{**\prime\prime}$ express the relations of *identity* and *being less than* respectively.

## System H

Axioms:

All wffs of Q that are axioms of QS by any of the schemata QS 1–7

All wffs of Q that are axioms of $QS^=$ by the schema $QS^=$ 2

Also:

1 $\bigwedge x[x = x]$ [i.e. $QS^=$ 1]
2 $\bigwedge x \bigwedge y \bigwedge z(x = y \supset (y = z \supset x = z))$
3 $\bigwedge x \bigwedge y \bigwedge z(x = y \supset (x = z \supset y = z))$
4 $\bigwedge x \bigwedge y(x = y \supset Sx = Sy)$
5 $\bigwedge x[0 < Sx]$
6 $\bigwedge x[\sim x < 0]$
7 $\bigwedge x \bigwedge y(x < y \supset Sx < Sy)$
8 $\bigwedge x \bigwedge y(x < y \supset \sim x = y)$
9 $\bigwedge x \bigwedge y(x < y \supset \sim y = x)$
10 $\bigwedge x \bigwedge y(x < y \supset \sim y < x)$
11 $\bigwedge x \bigwedge y(\sim x < y \supset (\sim y < x \supset x = y))$
12 $\bigwedge x \bigwedge y(y < Sx \supset (y < x \vee y = x))$
13 $\bigwedge x[x + 0 = x]$
14 $\bigwedge x \bigwedge y[x + Sy = S[x + y]]$
15 $\bigwedge x \bigwedge y \bigwedge z \bigwedge w(x = y \supset (z + x = w \supset z + y = w))$
16 $\bigwedge x \bigwedge y \bigwedge z \bigwedge w(x = y \supset (z = x + w \supset z = y + w))$
17 $\bigwedge x[x.0 = 0]$
18 $\bigwedge x \bigwedge y[x.Sy = [x.y] + x]$
19 $\bigwedge x \bigwedge y \bigwedge z \bigwedge w(x = y \supset (z.x = w \supset z.y = w))$
20 $\bigwedge x \bigwedge y \bigwedge z \bigwedge w(x = y \supset (z = x.w \supset z = y.w))$
21[1] $\bigwedge x[Px0 = S0]$
22[2] $\bigwedge x \bigwedge y[PxSy = [Pxy].x]$

[1] On the intended interpretation 21 means 'For every natural number $x$, $x^0 = 1$'.

[2] On the intended interpretation 22 means 'For every pair of natural numbers $x$ and $y$, $x^{y+1} = x^y.x$'.

23 $\bigwedge x \bigwedge y \bigwedge z \bigwedge w(x = y \supset (\mathrm{P}zx = w \supset \mathrm{P}zy = w))$
24 $\bigwedge x \bigwedge y \bigwedge z \bigwedge w(((y < x \wedge x = y + z) \vee (\sim y < x \wedge z = 0)) \supset$
$(((y < x \wedge x = y + w) \vee (\sim y < x \wedge w = 0)) \supset z = w))$

Our main purpose in formulating this set of axioms is to make the metatheoretical proofs in §56 as simple and short as possible. We have not made any attempt at economy, independence or elegance. We could instead have taken the seven axioms of Robinson's arithmetic (§48). These, added to the axioms of QS⁼, would have made an adequate system, and a much simpler and more elegant one. But the metatheoretical proofs would have been much more laborious.

*The consistency of H*

H satisfies the three conditions of §51 for being a formal system of arithmetic. Since it is a first order theory it also satisfies the third condition for respectability. The other two requirements for respectability are

1. That it should be consistent
2. That every decidable set of natural numbers should be represented in it

It seems clear to intuition that the intended interpretation of H is a model of H, and hence that H is consistent. But how can this be proved? It seems that any proof of the consistency of H would have to appeal to principles at least as open to question as those found in H. Our definition of a model, for example, makes use of set theory, and nobody has proved that set theory is consistent. So here we shall simply trust our intuition that the axioms of H do all come out true on the intended interpretation, and that therefore H is consistent. (For more on this cf. the references given in §59, under the heading *Gödel's second incompleteness theorem.*)

This leaves the second condition for respectability. We shall prove that every recursive set of natural numbers is represented in H, and then appeal to Church's Thesis in the form 'Every decidable set of natural numbers is recursive'.

## 56 Proof of the undecidability of H

Throughout this section:

$\vdash$ is an abbreviation for $\vdash_H$;

$\bar{m}, \bar{n}, \bar{k}, \overline{m+1}, \overline{m+k}, \overline{m+k+1}$ (e.g.)
are the numerals in H denoting the numbers
$m, n, k, m+1, m+k, m+k+1$ respectively.

We give proofs in compressed form. For example, we say simply 'By axiom 1' when in fact we mean 'By axiom 1, QS 4, and Modus Ponens'. Again, we shall not usually bother to cite tautologies but simply write 'By propositional logic'. We shall also omit some brackets. In virtue of the completeness theorem for QS we have at our disposal every logically valid formula of Q.

To get the undecidability of H we want to establish that every recursive function is definable in H.[1] We begin (56.1–56.8) with some preliminary metatheorems about H.

56.1 *For any natural numbers m, n, if* $m = n$ *then* $\vdash \bar{m} = \bar{n}$

*Proof.* By axiom 1.

56.2 *For any natural numbers m, n, if* $m < n$ *then* $\vdash \bar{m} < \bar{n}$

*Proof.* $m < n$ iff $0 < n - m$. Suppose $0 < n - m$. Then $\vdash 0 < \overline{n-m}$ by axiom 5. Then by $m$ applications of axiom 7 we get $\vdash \bar{m} < \overline{n-m+m}$, i.e. $\vdash \bar{m} < \bar{n}$.

56.3 *For any natural numbers m, n, if* $m \neq n$ *then* $\vdash \sim \bar{m} = \bar{n}$

*Proof.* Suppose $m \neq n$. Then either $m < n$ or $n < m$. So by 56.2 we have either $\vdash \bar{m} < \bar{n}$ or $\vdash \bar{n} < \bar{m}$. So by axiom 8 or axiom 9 $\vdash \sim \bar{m} = \bar{n}$.

56.4 *For any natural numbers m, n, if* $\sim m < n$ *then* $\vdash \sim \bar{m} < \bar{n}$

*Proof.* Suppose $\sim m < n$. Then either $n < m$ or $m = n$. So either $\vdash \bar{n} < \bar{m}$ by 56.2 or $\vdash \bar{m} = \bar{n}$ by 56.1. Then by either axiom 10 or axiom 8 and propositional logic $\vdash \sim \bar{m} < \bar{n}$.

[1] This part of the proof closely follows Grzegorczyk's proof that every recursive relation is strongly represented in his system Ar (*Fonctions récursives*, chap. v).

56.5 *For any natural number m,* $\vdash S\bar{m} = \overline{m+1}$

*Proof.* The numeral $S\bar{m}$ is the same numeral as $\overline{m+1}$. So by axiom 1 $\vdash S\bar{m} = \overline{m+1}$.

56.6 *For any natural numbers m, n,* $\vdash \bar{m}+\bar{n} = \overline{m+n}$

*Proof.* By induction in the metalanguage.

*Basis*: To prove $\vdash \bar{m}+\bar{0}+\overline{m+0}$

By axiom 13 $\bar{m}+0=\bar{m}$. But $\bar{m}$ is $\overline{m+0}$ and 0 is $\bar{0}$. So $\vdash \bar{m}+\bar{0} = \overline{m+0}$.

*Induction Step*

Assume $\vdash \bar{m}+\bar{k} = \overline{m+k}$. To prove $\vdash \bar{m}+\overline{k+1} = \overline{m+k+1}$.

$\vdash \bar{m}+S\bar{k} = S[\bar{m}+\bar{k}]$ by axiom 14 (1)

By the induction hypothesis $\vdash \bar{m}+\bar{k} = \overline{m+k}$ (2)

So from (2) by axiom 4 $\vdash S[\bar{m}+\bar{k}] = S\overline{m+k}$ (3)

So from (1) and (3) by axiom 2 $\vdash \bar{m}+S\bar{k} = S\overline{m+k}$ (4)

By 56.5 $\vdash S\overline{m+k} = \overline{m+k+1}$ (5)

So from (4) and (5) by axiom 2 $\vdash \bar{m}+S\bar{k} = \overline{m+k+1}$ (6)

By 56.5 $\vdash Sk = \overline{k+1}$ (7)

So from (7) and (6) by axiom 15 $\vdash \bar{m}+\overline{k+1} = \overline{m+k+1}$

This completes the Induction Step and the proof.

56.7 *For any natural numbers m, n* $\vdash \bar{m}.\bar{n} = \overline{m.n}$

*Proof.* By induction in the metalanguage.

*Basis*: To prove $\vdash \bar{m}.\bar{0} = \overline{m.0}$.

By axiom 17 $\vdash \bar{m}.0 = 0$. But 0 is $\bar{0}$ and $\overline{m.0}$ is $\bar{0}$.

So $\vdash \bar{m}.\bar{0} = \overline{m.0}$.

*Induction Step*

Assume $\vdash \bar{m}.\bar{k} = \overline{m.k}$. To prove $\vdash \bar{m}.\overline{k+1} = \overline{[m.k]+m}$.

By axiom 18 $\vdash \bar{m}.S\bar{k} = [\bar{m}.\bar{k}]+\bar{m}$ (1)

By the induction hypothesis $\vdash \bar{m}.\bar{k} = \overline{m.k}$ (2)

From (2) and (1) by axiom 16 $\vdash \bar{m}.S\bar{k} = \overline{m.k}+\bar{m}$ (3)

By 56.6 $\vdash \overline{m.k}+\bar{m} = \overline{[m.k]+m}$ (4)

So from (3) and (4) by axiom 2 $\vdash \bar{m}.S\bar{k} = \overline{[m.k]+m}$ (5)

But from 56.5 $\vdash S\bar{k} = \overline{k+1}$ (6)

So from (6) and (5) by axiom 19 $\vdash \bar{m}.\overline{k+1} = \overline{[m.k]+m}$

This completes the Induction Step and the proof.

56.8 *For any natural numbers m, n* $\vdash P\bar{m}\,\bar{n} = \overline{m^n}$

*Proof.* By induction.

*Basis*: To prove $\vdash P\bar{m}\,\bar{0} = \overline{m^0}$.

By axiom 21 $\vdash P\bar{m}0 = S0$

But 0 is $\bar{0}$ and $\overline{m^0}$ is S0. So $\vdash P\bar{m}0 = \overline{m^0}$.

*Induction Step*

Assume $\vdash P\bar{m}\,\bar{k} = \overline{m^k}$. To prove $\vdash P\bar{m}\,\overline{k+1} = \overline{m^{k+1}}$.

By axiom 22 $\vdash P\bar{m}S\bar{k} = [P\bar{m}\,\bar{k}].\bar{m}$ .. .. .. (1)

By the induction hypothesis $\vdash P\bar{m}\,\bar{k} = \overline{m^k}$ .. .. .. (2)

So from (2) and (1) by axiom 20 $\vdash P\bar{m}S\bar{k} = \overline{m^k}.\bar{m}$ .. (3)

By 56.7 $\vdash \overline{m^k}.\bar{m} = \overline{m^k.m}$ .. .. .. .. .. (4)

So from (3) and (4) by axiom 2 $\vdash P\bar{m}S\bar{k} = \overline{m^k.m}$ .. .. (5)

But by 56.5 $\vdash S\bar{k} = \overline{k+1}$ .. .. .. .. .. (6)

So from (6) and (5) by axiom 23 $\vdash P\bar{m}\,\overline{k+1} = \overline{m^k.m}$

But $\overline{m^k.m}$ is $\overline{m^{k+1}}$.

So $\vdash P\bar{m}\,\overline{k+1} = \overline{m^{k+1}}$.

This completes the Induction Step and the proof.

*Definability in H of the initial functions (cf. pp. 235-6)*

56.9 *The successor function is definable in H by the formula* $Sx = y$

*Proof*

Condition (i). Suppose $m+1 = n$. Then $\overline{m+1}$ is $\bar{n}$. So by axiom 1 $\vdash \overline{m+1} = \bar{n}$. But $\overline{m+1}$ is $S\bar{m}$. So $\vdash S\bar{m} = \bar{n}$. So if $m+1 = n$, then $\vdash S\bar{m} = \bar{n}$.

Condition (ii). Suppose $m+1 \neq n$. Then by 56.3 $\vdash \sim\overline{m+1} = \bar{n}$, i.e. $\vdash \sim S\bar{m} = \bar{n}$. So if $m+1 \neq n$, then $\vdash \sim S\bar{m} = \bar{n}$.

Condition (iii). By axiom 3 $\vdash (Sx = y \supset (Sx = z \supset y = z))$. Then by 45.5 $\vdash \Lambda x \Lambda y \Lambda z(Sx = y \supset (Sx = z \supset y = z))$.

56.10 *The sum function is definable in H by the formula* $x + y = z$

*Proof*

Condition (i). Suppose $m + n = k$. Then $\overline{m+n}$ is $\bar{k}$. So by axiom 1 $\vdash \overline{m+n} = \bar{k}$. So, since by 56.6 $\vdash \bar{m} + \bar{n} = \overline{m+n}$, $\vdash \bar{m} + \bar{n} = \bar{k}$, by axiom 2.

Condition (ii). Suppose $m + n \neq k$. Then by 56.3 $\vdash \sim\overline{m+n} = \bar{k}$.

Then, since $\vdash \bar{m} + \bar{n} = \overline{m+n}$, $\vdash \sim \bar{m} + \bar{n} = \bar{k}$, by axiom 3 and propositional logic. So if $m + n \neq k$, then $\vdash \sim \bar{m} + \bar{n} = \bar{k}$.

Condition (iii). By axiom 3 and 45.5
$\vdash \Lambda x \Lambda y \Lambda z \Lambda w(x + y = z \supset (x + y = w \supset z = w))$.

56.11 *The product function is definable in H by the formula* $x.y = z$

*Proof*

Condition (i). Suppose $m.n = k$. Then $\vdash \overline{m.n} = \bar{k}$, by axiom 1. So since by 56.7 $\vdash \bar{m}.\bar{n} = \overline{m.n}$, $\vdash \bar{m}.\bar{n} = \bar{k}$, by axiom 2. So if $m.n = k$, then $\vdash \bar{m}.\bar{n} = \bar{k}$.

Condition (ii). Suppose $m.n \neq k$. Then $\vdash \sim \overline{m.n} = \bar{k}$. Then, since $\vdash \bar{m}.\bar{n} = \overline{m.n}$, $\vdash \sim \bar{m}.\bar{n} = \bar{k}$, by axiom 3 and propositional logic. So if $m.n \neq k$, then $\vdash \sim \bar{m}.\bar{n} = \bar{k}$.

Condition (iii). By axiom 3 and 45.5
$\vdash \Lambda x \Lambda y \Lambda z \Lambda w(x.y = z \supset (x.y = w \supset z = w))$.

56.12 *The power function is definable in H by the formula* $Pxy = z$

*Proof*

Condition (i). Suppose $m^n = k$. Then $\vdash \overline{m^n} = \bar{k}$, by axiom 1. So, since $\vdash \mathrm{P}\,\bar{m}\,\bar{n} = \overline{m^n}$ (56.8), $\vdash \mathrm{P}\,\bar{m}\,\bar{n} = \bar{k}$, by axiom 2. So if $m^n = k$, then $\vdash \mathrm{P}\,\bar{m}\,\bar{n} = \bar{k}$.

Condition (ii). Suppose $m^n \neq k$. Then $\vdash \sim \overline{m^n} = \bar{k}$. Then since $\vdash \mathrm{P}\,\bar{m}\,\bar{n} = \overline{m^n}$, $\vdash \sim P\,\bar{m}\,\bar{n} = \bar{k}$, by axiom 3 and propositional logic. So if $m^n \neq k$, then $\vdash \sim \mathrm{P}\,\bar{m}\,\bar{n} = \bar{k}$.

Condition (iii). By axiom 3 and 45.5
$\vdash \Lambda x \Lambda y \Lambda z \Lambda w(\mathrm{P}xy = z \supset (\mathrm{P}xy = w \supset z = w))$.

56.13 *The arithmetic difference function is definable in H by the formula* $(y < x \wedge x = y + z) \vee (\sim y < x \wedge z = 0)$

*Proof*

Condition (i). Suppose $m \dot{-} n = k$. There are two possible cases: (1) $n < m$ (2) $m \leqslant n$.

1. Suppose $n < m$. Then $m \dot{-} n = m - n$. So $m = n + (m \dot{-} n)$. Then we have $\vdash \bar{n} < \bar{m}$ and $\vdash \bar{m} = \overline{n + (m \dot{-} n)}$ and so by 56.6 $\vdash \bar{m} = \bar{n} + \overline{m \dot{-} n}$. So $\vdash \bar{n} < \bar{m} \wedge \bar{m} = \bar{n} + \overline{m \dot{-} n}$. But $\overline{m \dot{-} n}$ is $\bar{k}$. So $\vdash \bar{n} < \bar{m} \wedge \bar{m} = \bar{n} + \bar{k}$.

2. Suppose $m \leqslant n$. Then $m \dot{-} n = 0$. And we have $\vdash \sim \bar{n} < \bar{m} \wedge$

$\overline{m \dot{-} n} = 0$. But $\overline{m \dot{-} n}$ is $\bar{k}$. So $\vdash \sim \bar{n} < \bar{m} \wedge \bar{k} = 0$.
So we have:

If $m \dot{-} n = k$ and $n < m$, then $\vdash \bar{n} < \bar{m} \wedge \bar{m} = \bar{n} + \bar{k}$.

If $m \dot{-} n = k$ and $m \leqslant n$, then $\vdash \sim \bar{n} < \bar{m} \wedge \bar{k} = 0$.

So if $m \div n = k$,then either $\vdash \bar{n} < \bar{m} \wedge \bar{m} = \bar{n} + \bar{k}$ or $\vdash \sim \bar{n} < \bar{m} \wedge \bar{k} = 0$.

Now if either $\vdash$ A or $\vdash$ B, then either $\vdash$ A $\vee$ C or $\vdash$ B $\vee$ D where C and D are arbitrary wffs. So if either $\vdash$ A or $\vdash$ B then either $\vdash$ A $\vee$ B or $\vdash$ B $\vee$ A. But $\vdash$ (B $\vee$ A) $\supset$ (A $\vee$ B). So if either $\vdash$ A or $\vdash$ B then either $\vdash$ A $\vee$ B or $\vdash$ A $\vee$ B; i.e. if either $\vdash$ A or $\vdash$ B then $\vdash$ A $\vee$ B.

So if $m \dot{-} n = k$ then $\vdash (\bar{n} < \bar{m} \wedge \bar{m} = \bar{n} + \bar{k}) \vee (\sim \bar{n} < \bar{m} \wedge \bar{k} = 0)$.

Condition (ii). Suppose $m \div n \neq k$. Again there are two possible cases: (1) $n < m$ (2) $m \leqslant n$.

1. Suppose $n < m$. Then, as in (i) 1 above, we have $m = n + (m \dot{-} n)$ and therefore, since $m \dot{-} n \neq k$, $m \neq n + k$. So we have $\vdash \sim \bar{m} = \bar{n} + \bar{k}$. But we also have $\vdash \bar{n} < \bar{m}$. So $\vdash \bar{n} < \bar{m} \wedge \sim \bar{m} = \bar{n} + \bar{k}$.

2. Suppose $m \leqslant n$. Then $m \dot{-} n = 0$. But $m \dot{-} n \neq k$. So $k \neq 0$. So $\vdash \sim \bar{k} = 0$. But also $\vdash \sim \bar{n} < \bar{m}$. So we have $\vdash \sim \bar{n} < \bar{m} \wedge \sim \bar{k} = 0$. So, as for (i) above, we have:

If $m \dot{-} n \neq k$ then $\vdash (\bar{n} < \bar{m} \wedge \sim \bar{m} = \bar{n} + \bar{k}) \vee (\sim \bar{n} < \bar{m} \wedge \sim \bar{k} = 0)$.

From this by propositional logic we get:

If $m \dot{-} n \neq k$ then $\vdash \sim ((\bar{n} < \bar{m} \wedge \bar{m} = \bar{n} + \bar{k}) \vee (\sim \bar{n} < \bar{m} \wedge \bar{k} = 0))$.

Condition (iii). By axiom 24.

Next we want to show that any function obtained by either the $\mu$-operation or composition from functions definable in H is also definable in H. We prove a preliminary Lemma.

56.14 *Lemma: Let S be a first order theory in which the identity function is strongly represented by '$x = y$'. Let $A(x, y)$ be a formula of S of which it is true that*

$$\vdash_S \bigwedge x \bigwedge y \bigwedge z (A(x, y) \supset (A(x, z) \supset y = z))$$

*Then in order to show that a function f of one argument is definable in S by the formula $A(x, y)$, it is enough to show that* $\vdash A(\bar{m}, \overline{f(m)})$

*Proof*

Condition (i). Suppose $f(m) = n$. Then $\overline{f(m)}$ is $\bar{n}$. So if we can show that $\vdash_S A(\bar{m}, \overline{f(m)})$, we shall have shown that $\vdash_S A(\bar{m}, \bar{n})$, and so that condition (i) for definability is satisfied.

Condition (ii). Suppose $f(m) \neq n$. Then, since *ex hypothesi* identity is strongly represented in S by '$x = y$', $\vdash_S \sim \overline{f(m)} = \bar{n}$. Also *ex hypothesi* we have $\vdash_S A(\bar{m}, \overline{f(m)}) \supset (A(\bar{m}, \bar{n}) \supset \overline{f(m)} = \bar{n})$. So by propositional logic $\vdash_S A(\bar{m}, \overline{f(m)}) \supset (\sim \overline{f(m)} = \bar{n} \supset \sim A(\bar{m}, \bar{n}))$. So if we can show that $\vdash_S A(\bar{m}, \overline{f(m)})$, by applying Modus Ponens twice we shall get $\vdash_S \sim A(\bar{m}, \bar{n})$, and so condition (ii) will be satisfied.

Condition (iii). This is satisfied *ex hypothesi.*

The Lemma can be generalised to functions of more than one argument.

*Composition*

For simplicity we consider only functions of one argument. The proof can easily be generalised to cover functions of more than one argument.

Let $f$ be a function definable in H by a formula $A(x, y)$ and $g$ a function definable in H by a formula $B(x, y)$. We shall show that the function $h$ defined by the rule

$$h(x) = f(g(x))$$

is definable in H by the formula

$$\bigvee w(B(x, w) \wedge A(w, y))$$

*Condition (iii)*

*Ex hypothesi* we have

$\vdash \bigwedge x \bigwedge y \bigwedge z(A(x, y) \supset (A(x, z) \supset y = z))$

and

$\vdash \bigwedge x \bigwedge y \bigwedge z(B(x, y) \supset (B(x, z) \supset y = z))$

We want to show that

$\vdash \bigwedge x \bigwedge y \bigwedge z(\bigvee w(B(x, w) \wedge A(w, y)) \supset (\bigvee w(B(x, w) \wedge A(w, z)) \supset y = z))$

Let P, Q, R, S be abbreviations for the formulas shown:

P: $\bigwedge x \bigwedge y \bigwedge z(A(x, y) \supset (A(x, z) \supset y = z))$

Q: $\bigwedge x \bigwedge y \bigwedge z(B(x, y) \supset (B(x, z) \supset y = z))$

R: $B(x, w) \wedge A(w, y)$

S: $B(x, w) \wedge A(w, z)$

Then we want to show that, given $\vdash$ P and $\vdash$ Q,

$\vdash \bigwedge x \bigwedge y \bigwedge z(\bigvee w R \supset (\bigvee w S \supset y = z))$

Let H′ be the first order theory that results from adding denumerably many new constants $b'$, $b''$, $b'''$, . . . to H. And let H″ be the first order theory that results from adding denumerably many new constants $c', c'', c''', \ldots$ to H$'$. It is obvious that

(1) $\mathrm{R}b'/w, \bigvee w\mathrm{S} \vdash_{\mathrm{H}'} \bigvee w\mathrm{S}$

We shall show below that

(2) $\mathrm{R}b'/w, \bigvee w\mathrm{S}, \mathrm{S}c'/w \vdash_{\mathrm{H}''} y = z$

From (1) and (2) we have, in virtue of Rule C (45.24 p. 192),

(3) $\mathrm{R}b'/w, \bigvee w\mathrm{S} \vdash_{\mathrm{H}'} y = z$

From (3) it is obvious that

(4) $\bigvee w\mathrm{R}, \bigvee w\mathrm{S}, \mathrm{R}b'/w \vdash_{\mathrm{H}'} y = z$

It is also obvious that

(5) $\bigvee w\mathrm{R}, \bigvee w\mathrm{S} \vdash_{\mathrm{H}} \bigvee w\mathrm{R}$

From (4) and (5), in virtue of Rule C,

(6) $\bigvee w\mathrm{R}, \bigvee w\mathrm{S} \vdash_{\mathrm{H}} y = z$

From (6) by the Deduction Theorem

(7) $\vdash_{\mathrm{H}} \bigvee w\mathrm{R} \supset (\bigvee w\mathrm{S} \supset y = z)$

Then by 45.4, applied three times,

(8) $\vdash_{\mathrm{H}} \bigwedge x \bigwedge y \bigwedge z(\bigvee w\mathrm{R} \supset (\bigvee w\mathrm{S} \supset y = z))$

which is what we want.

We still have to show that (2) holds:

1. $\vdash_{\mathrm{H}''} \mathrm{B}(x, b') \supset (\mathrm{B}(x, c') \supset b' = c')$ [From $\vdash$ Q by repeated uses of K4 and MP]
2. $\mathrm{R}b'/w, \mathrm{S}c'/w$ [i.e. $\mathrm{B}(x, b') \wedge \mathrm{A}(b', y), \mathrm{B}(x, c') \wedge \mathrm{A}(c', z)$: abbreviate this to T] $\vdash_{\mathrm{H}''} \mathrm{B}(x, b') \wedge \mathrm{A}(b', y)$
3. $\mathrm{T} \vdash_{\mathrm{H}''} \mathrm{B}(x, b')$ [From 2 by propositional logic]
4. $\mathrm{T} \vdash_{\mathrm{H}''} \mathrm{B}(x, c') \supset b' = c'$ [1, 3, MP]
5. $\mathrm{T} \vdash_{\mathrm{H}''} \mathrm{B}(x, c') \wedge \mathrm{A}(c', z)$ [i.e. $\mathrm{S}c'/w$]
6. $\mathrm{T} \vdash_{\mathrm{H}''} \mathrm{B}(x, c')$ [From 5 by propositional logic]
7. $\mathrm{T} \vdash_{\mathrm{H}''} b' = c'$ [4, 6, MP]
8. $\vdash_{\mathrm{H}''} b' = c' \supset (\mathrm{A}(b', y) \supset \mathrm{A}(c', y))$ [QS = 2]
9. $\mathrm{T} \vdash_{\mathrm{H}''} \mathrm{A}(b', y) \supset \mathrm{A}(c', y)$ [7, 8, MP]
10. $\mathrm{T} \vdash_{\mathrm{H}''} \mathrm{A}(b', y)$ [From 2 by propositional logic]
11. $\mathrm{T} \vdash_{\mathrm{H}''} \mathrm{A}(c', y)$ [9, 10, MP]
12. $\vdash_{\mathrm{H}''} \mathrm{A}(c', y) \supset (\mathrm{A}(c', z) \supset y = z)$ [From $\vdash$ P by repeated uses of K4 and MP]
13. $\mathrm{T} \vdash_{\mathrm{H}''} \mathrm{A}(c', z) \supset y = z$ [11, 12, MP]
14. $\mathrm{T} \vdash_{\mathrm{H}''} \mathrm{A}(c', z)$ [From 5 by propositional logic]
15. $\mathrm{T} \vdash_{\mathrm{H}''} y = z$ [13, 14, MP]

So $\mathrm{T}, \bigvee w\mathrm{S} \vdash_{\mathrm{H}''} y = z$.

So $\mathrm{R}b'/w, \bigvee w\mathrm{S}, \mathrm{S}c'/w \vdash_{\mathrm{H}''} y = z$, which is what we want.

So condition (iii) is satisfied. So by the Lemma 56.14 in order to prove that $h$ is definable in H we need now only show that

$$\vdash \bigvee w(\mathrm{B}(\bar{m}, w) \wedge \mathrm{A}(w, \overline{f(g(m))}))$$

Since $g$ is definable by $\mathrm{B}(x, y)$ we have

$$\vdash \mathrm{B}(\bar{m}, \overline{g(m)})$$

Since $f$ is definable by $\mathrm{A}(x, y)$ we have

$$\vdash \mathrm{A}(\overline{g(m)}, \overline{f(g(m))})$$

So we have

$$\vdash \mathrm{B}(\bar{m}, \overline{g(m)}) \wedge \mathrm{A}(\overline{g(m)}, \overline{f(g(m))})$$

Hence by 45.25 [*If $t$ is a closed term of K, then* $\vdash_K At/v \supset \bigvee vA$] and MP

$$\vdash \bigvee w(\mathrm{B}(\bar{m}, w) \wedge \mathrm{A}(w, \overline{f(g(m))}))$$

which is what we want.

Before considering the $\mu$-operation we prove a Lemma.

56.15 $\vdash x < \bar{k} \supset (x = 0 \vee \ldots \vee x = \overline{k-1})$

*Proof*

*Basis*: $\bar{k} = 0$

In virtue of the tautology $\sim \mathrm{A} \supset (\mathrm{A} \supset \mathrm{B})$

$\vdash \sim x < 0 \supset (x < 0 \supset (x = 0 \vee \ldots \vee x = \overline{k-1}))$.

By axiom 6

$\vdash \sim x < 0$.

Therefore $\vdash x < 0 \supset (x = 0 \vee \ldots \vee x = \overline{k-1})$.

*Induction Step*

Assume the Lemma holds for all $\bar{k}$ where $k \leqslant m$.

To prove $\vdash x < \overline{m+1} \supset (x = 0 \vee \ldots \vee x = \bar{m})$.

By axiom 12 $x < \overline{m+1} \vdash x < \bar{m} \vee x = \bar{m}$.

By the induction hypothesis $\vdash x < \bar{m} \supset (x = 0 \vee \ldots \vee x = \overline{m-1})$.

$\vdash (\mathrm{A} \vee \mathrm{B}) \supset ((\mathrm{A} \supset \mathrm{C}) \supset (\mathrm{C} \vee \mathrm{B}))$.

So $x < \overline{m+1} \vdash x = 0 \vee \ldots \vee x = \overline{m-1} \vee x = \bar{m}$.

So by the Deduction Theorem $\vdash x < \overline{m+1} \supset (x = 0 \vee \ldots \vee x = \bar{m})$.

*The $\mu$-operation*

For simplicity we consider only functions of two arguments.

Let $f$ be a function of two arguments definable in H by the formula $\mathrm{A}(x, y, z)$ and satisfying the condition that for each natural number $m$ there is a natural number $n$ such that $f(m, n) = 0$.

We shall show that the function $g$ defined by the rule

$$g(m) = \mu n\{f(m, n) = 0\}$$

is definable in H by the formula

$$A(x, y, 0) \wedge \bigwedge z(z < y \supset \sim A(x, z, 0))$$

We abbreviate this formula to $B(x, y)$.

*Condition (iii)*

We want to show that $\vdash B(x, y) \supset (B(x, z) \supset y = z)$.

$B(x, y)$ is $A(x, y, 0) \wedge \bigwedge z(z < y \supset \sim A(x, z, 0))$.

$B(x, z)$ is $A(x, z, 0) \wedge \bigwedge y(y < z \supset \sim A(x, y, 0))$.

We have:

$B(x, y), B(x, z), z < y \vdash A(x, z, 0) \wedge \sim A(x, z, 0)$

and

$B(x, y), B(x, z), y < z \vdash A(x, y, 0) \wedge \sim A(x, y, 0))$

So $B(x, y), B(x, z) \vdash \sim z < y$ and $B(x, y), B(x, z \vdash \sim y < z$.

So by axiom 11

$B(x, y), B(x, z) \vdash y = z$.

So $\vdash B(x, y) \supset (B(x, z) \supset y = z)$.

By the Lemma 56.14 it only remains to show that $\vdash B(\bar{m}, \overline{g(m)})$, i.e. that $\vdash A(\bar{m}, \overline{g(m)}, 0) \wedge \bigwedge z(z < \overline{g(m)} \supset \sim A(\bar{m}, z, 0))$.

Since *ex hypothesi* the function $f$ is definable by the formula $A(x, y, z)$, we have for every pair of natural numbers $m, n$

$$\vdash A(\bar{m}, \bar{n}, \overline{f(m, n)}) \qquad (1)$$

We also have *ex hypothesi*

For each natural number $m$ there is a natural number $n$ such that $f(m, n) = 0$ (2)

and by definition

$$g(m) = \mu n\{f(m, n) = 0\} \qquad (3)$$

From 2 and 3 we have

$$f(m, g(m)) = 0 \text{ for every natural number } m \qquad (4)$$

From 1 and 4 we have

$$\vdash A(\bar{m}, \overline{g(m)}, 0) \qquad (5)$$

which is the first conjunct of what we want to prove.

From 3 we have

If $n < g(m)$ then $f(m, n) \neq 0$. (6)

Since $f$ is definable by $A(x, y, z)$ and condition (ii) of definability ensures that if $f(m, n) \neq 0$ then $\vdash \sim A(\bar{m}, \bar{n}, 0)$, from 6 we get

If $n < g(m)$ then $\vdash \sim A(\bar{m}, \bar{n}, 0)$ (7)

1. Now suppose $g(m) = 0$. Then by axiom 6 $\vdash \sim z < \overline{g(m)}$. So by propositional logic $\vdash z < \overline{g(m)} \supset \sim A(m, z, 0)$. So by 45.4 $\vdash \bigwedge z(z < \overline{g(m)} \supset \sim A(\bar{m}, z, 0))$. So:

If $g(m) = 0$, then $\vdash \bigwedge z(z < \overline{g(m)} \supset \sim A(\bar{m}, z, 0))$ (8)

2. Suppose $g(m) > 0$. By QS⁼2 $\vdash z = 0 \supset (\sim A(\bar{m}, 0, 0) \supset \sim A(\bar{m}, z, 0))$. So by propositional logic $\vdash \sim A(\bar{m}, 0, 0) \supset (z = 0 \supset \sim A(\bar{m}, z, 0))$. Since *ex hypothesi* $0 < g(m)$, from 7 we get $\vdash \sim A(\bar{m}, 0, 0)$. So by Modus Ponens $\vdash z = 0 \supset \sim A(\bar{m}, z, 0)$. Similarly for each $n < g(m)$. So:

$$\vdash (z = 0 \supset \sim A(\bar{m}, z, 0)) \wedge \ldots \wedge (z = \overline{g(m) - 1} \supset \sim A(\bar{m}, z, 0)).$$

By propositional logic

$$\vdash ((z = 0 \supset \sim A(\bar{m}, z, 0)) \wedge \ldots \wedge (z = \overline{g(m) - 1} \supset \sim A(\bar{m}, z, 0))) \supset ((z = 0 \vee \ldots \vee z = \overline{g(m) - 1}) \supset \sim A(\bar{m}, z, 0)).$$

So by Modus Ponens

$$\vdash (z = 0 \vee \ldots \vee z = \overline{g(m) - 1}) \supset \sim A(\bar{m}, z, 0) \quad (9)$$

From the Lemma 56.15

$$\vdash z < \overline{g(m)} \supset (z = 0 \vee \ldots \vee z = \overline{g(m) - 1}) \quad (10)$$

From 9 and 10

$$\vdash z < \overline{g(m)} \supset \sim A(\bar{m}, z, 0)$$

So by 45.4

$$\vdash \bigwedge z(z < \overline{g(m)} \supset \sim A(\bar{m}, z, 0)).$$

So:

If $g(m) > 0$, then $\vdash \bigwedge z(z < \overline{g(m)} \supset \sim A(\bar{m}, z, 0))$ (11)

From 8 and 11

$$\vdash \bigwedge z(z < \overline{g(m)} \supset \sim A(\bar{m}, z, 0)) \quad (12)$$

From 5 and 12 and propositional logic

$$\vdash A(\bar{m}, \overline{g(m)}, 0) \wedge \bigwedge z(z < \overline{g(m)} \supset \sim A(\bar{m}, z, 0))$$

Q.E.D.

To sum up:

56.16 *Every recursive function is definable in H*

So, given the consistency of H (§55) and the remarks in §54:

56.17 *Every recursive function is strongly represented in H*

56.18 *Every recursive function is represented in H*

56.19 *Every recursive set of natural numbers is represented in H*

Hence, assuming Church's Thesis:

56.20 *Every decidable set of natural numbers is represented in H*

So H is a respectable formal system of arithmetic.[1] So by 51.10

56.21 *H is undecidable*

*Note.* This result is under the assumption of Church's Thesis. (Also, it could be objected that we have not proved that H is consistent.)

## 57 Proof of the undecidability of QS$^=$. The undecidability of QS

H is a finite extension of QS$^=$ [the axioms of H are the axioms of QS$^=$ and the axioms numbered 2–24]. So by 51.14 and 56.21:

57.1 *QS$^=$ is undecidable*

*Note.* This is still under the assumption of Church's Thesis.

H is not a finite extension of QS, since H includes among its axioms all the infinitely many axioms by the axiom-schema QS$^=$ 2, and these are not axioms of QS.

We appealed to axioms by QS$^=$ 2 at two places in the proof that every recursive function is definable in H [pp. 245 and 248]. If we could drop QS$^=$ 2 from H and replace it by some finite set of axioms and still prove that in the resulting system every recursive function was definable, or, alternatively, if we could show, e.g. by induction on the number of function letters in an arbitrary formula, that even when QS$^=$ 2 is dropped the general

[1] Given that it satisfies the other conditions for respectability, of course.

substitutivity of identity is provable from the remaining axioms of H, then we should be able to prove from 51.14 the undecidability of QS. But there is no painless way of doing this. So we content ourselves with proving the undecidability of $QS^=$, and merely state without proof:

[57.2] *QS is undecidable*

Also:

[57.3] *The set of logically valid wffs of Q not containing function symbols or individual constants is undecidable*

A formal system having exactly that set of wffs as theorems is sometimes called a 'pure predicate calculus'. [57.3] says that such a calculus is undecidable.

57.4 *The set of theorems of $QS^=$ [QS] is effectively enumerable. But the set of formulas of Q that are not theorems of $QS^=$ [QS] is not effectively enumerable*

*Proof*

1. The set of formulas of $QS^=$ is decidable and there is an effective method for telling whether or not something is a proof in $QS^=$. So by 51.11 the theorems of $QS^=$ are effectively enumerable. Similarly for QS.

2. Arbitrary strings of symbols of Q can be effectively enumerated in the following way:

First we take all strings consisting of just one symbol. There are twelve of these, viz. $p$, $'$, $x$, $a$, $f$, $F$, $*$, $\sim$, $\supset$, $\wedge$, (, ). From now on let those twelve symbols be ordered lexicographically in that order.

We take all strings consisting of just two symbols, in lexicographical order, i.e. in the order

$$pp, p', px, pa, \ldots, 'p, '', 'x, \ldots, )\wedge, )(, )).$$

Then all strings consisting of just three symbols, just four symbols, just five symbols, and so on.

Let the first *formula* that turns up in this enumeration of arbitrary strings be assigned the number 0 [the first formula is in fact $p'$]. Let the second formula that turns up be assigned the number 1, the third the number 2, and so on. Every formula will turn up somewhere in the enumeration of the strings, so every formula will be assigned a natural number, and each distinct natural number will be assigned to a distinct formula.

Since $QS^=$ is not decidable, the set of theorems of $QS^=$ is not decidable. So the set of numbers assigned by our numbering to the theorems of $QS^=$ is not decidable either. But it is effectively enumerable, since the theorems of $QS^=$ are effectively enumerable and the numbering is effective. So by 51.5 the relative complement of this set is not effectively enumerable; i.e. the set of numbers of formulas of Q that are not theorems of $QS^=$ is not effectively enumerable. So the set of formulas of Q that are not theorems of $QS^=$ is not effectively enumerable.

[Similarly for QS.]

Informally, this means that it is possible to program a computer to churn out one by one the theorems of $QS^=$ [QS], without churning out any non-theorems, and in such a way that any formula that is a theorem will sooner or later (neglecting mechanical failure, end of the world, etc.) be churned out. But (says 57.4) it is not possible to do the same thing for the formulas that are not theorems. There is no way of programming a computer to churn out exclusively formulas that are not theorems in such a way that any formula that is not a theorem will sooner or later be churned out.

## 58 Decidable subclasses of logically valid formulas of Q. Prenex normal form. Skolem normal form. Two negative results

Though QS and $QS^=$ are undecidable, for particular subclasses of formulas of Q there are effective methods for telling whether or not a formula in the class is logically valid (or, equivalently, whether or not a formula in the class is a theorem of QS). For example, the class of logically valid formulas of Q containing only propositional symbols and connectives is decidable, by the truth-table method. Another decidable class is the class of logically valid formulas of Q containing no function symbols and whose only predicate symbols are one-place predicate symbols: we proved in §50 that this class is decidable.

Many of the decidable subclasses are defined in terms of prenex normal form, which we now explain.

*Prenex normal form*

If $\bigvee$ [the existential quantifier] were a symbol of Q we could give a simple definition of prenex normal form as follows (Definition I):

A formula A is in prenex normal form iff it is of the form $Qv_1 \ldots Qv_n$ B, where each Q is either $\bigwedge$ or $\bigvee$, $n \geqslant 0$, B is a wff, and no quantifiers occur in B.

The B in that definition is called the *matrix* of A, and the part of A (if any) that precedes the matrix is called the *prefix*.

Intuitively, a formula is in prenex normal form iff all the quantifiers occur unnegated in a row at the beginning of the formula and the scope of each quantifier extends to the end of the formula.

Since $\bigvee$ is not a symbol of Q we adopt the following definition instead (Definition II):

A formula of Q is in prenex normal form iff when each string of symbols in it of the form $\sim\bigwedge v\sim$ is replaced by $\bigvee v$, the result is a formula that is in prenex normal form according to Definition I.

Examples: The following formulas are in prenex normal form (allowing for the usual abbreviations):

1. $\bigwedge x \sim \bigwedge y \sim (Fx \supset Gy)$
2. $\bigwedge x \bigwedge y \sim \bigwedge z \sim \bigwedge w((p' \supset Fxy) \supset \sim Fzy)$
3. $Fxy \supset Fzy$
4. $p' \supset p''$

The following are *not* in prenex normal form:

5. $\bigwedge x \sim \sim \bigwedge y \sim (Fx \supset Gy)$
6. $(\bigwedge x \bigwedge y \sim \bigwedge z \sim \bigwedge w(p' \supset Fxy) \supset \sim Fzy)$

   [The scopes of the quantifiers do not extend to the end of the formula: or, equivalently, the part of the formula that follows the quantifiers, viz.

   $$(p' \supset Fxy) \supset \sim Fzy)$$

   is not a wff.] [*Note*. Outermost brackets have *not* been dropped in 6.]

7. $Fxy \supset \bigwedge z Fzy$
8. $\sim \bigwedge x(p' \supset p'')$

*Note*. Some definitions of prenex normal form require the

variables in the prefix to be distinct, and some require the matrix to contain, for each quantifier in the prefix, at least one occurrence of the variable that it binds. These requirements are not necessary for our present purpose.

For each of the following classes of formulas of Q there is an effective method for telling whether or not an arbitrary formula in the class is logically valid (or, equivalently, whether or not it is a theorem of QS). [In the descriptions that follow, 'existential quantifier' means $\sim\Lambda\ldots\sim$, and 'universal quantifier' means 'universal quantifier other than one that occurs as part of an existential quantifier'.]

[58.1] *Formulas, without function symbols, in prenex normal form in which the prefix contains*

(*a*) *no quantifiers,*[1] *or*

(*b*) *no existential quantifiers,*[1] *or*

(*c*) *only existential quantifiers,*[2] *or*

(*d*) *no existential quantifier in front of a universal quantifier,*[3] *or*

(*e*) *not more than one existential quantifier, or*

(*f*) *not more than two existential quantifiers and those not separated by any universal quantifier*

Results (*a*)–(*d*) are due to Bernays and Schönfinkel (1927), (*e*) to Ackermann (1928) and Skolem (1928), independently, (*f*) to Kalmár (1932), Gödel (1933), and Schütte (1933), all independently.

[58.2] *Formulas in prenex normal form in which the matrix is a disjunction of atomic wffs and/or the negations of atomic wffs* [Herbrand, 1929, pub. 1930]

There are other decidable subclasses as well, but their descriptions are more complicated. For further details see

[1] If the prefix contains no quantifiers or no existential quantifiers, then the formula is logically valid iff it is $k$-valid, where $k$ is the number of distinct variables in the matrix.

[2] If the prefix contains only existential quantifiers, then the formula is logically valid iff it is 1-valid.

[3] If all the universal quantifiers precede all the existential quantifiers, then the formula is logically valid iff it is $k$-valid, where $k$ is the number of distinct variables in the matrix that are bound by universal quantifiers.

Hilbert and Ackermann, *Principles of Mathematical Logic*, §12 or Church (1956, §46) or Ackermann (1954).

[58.1] and [58.2] have a wider application than might be expected, in virtue of the following theorem, which we state without proof:

[58.3] *There is an effective method for finding, for any formula A of Q, a formula A* of Q in prenex normal form such that* $\vdash_{QS} A \supset A^*$ and $\vdash_{QS} A^* \supset A$, *and therefore such that* $\vdash_{QS} A$ *iff* $\vdash_{QS} A^*$. [*Since* $\vdash_{QS} A$ *iff* $\models A$, *we also have* '. . . *such that* $\models A \supset A^*$ *and* $\models A^* \supset A$.]

*Skolem normal form*

A formula is in Skolem normal form iff it contains no function symbols and it is in prenex normal form and all existential quantifiers precede all universal quantifiers (of course not counting those universal quantifiers that occur as parts of existential quantifiers).

[58.4] *If A is a wff of Q without function symbols or individual constants, then there is an effective method for finding a wff B of Q in Skolem normal form such that* $\vdash_{QS} A$ *iff* $\vdash_{QS} B$ [*or*: '. . . *such that* $\models A$ *iff* $\models B$'] (N.B. But in contrast to [58.3] we do not necessarily have $\vdash_{QS} A \equiv B$.) [Skolem, 1919]

We end this section with two negative results.

By [57.3] the set of logically valid wffs of Q that have no function symbols or individual constants is undecidable. From [57.3] and [58.4] it follows that:

[58.5] *The set of logically valid wffs in Skolem normal form is not decidable*

Of course particular subsets of this set are decidable: cf. [58.1].

There is an even stronger result:

[58.6] *The set of logically valid wffs of Q containing only binary predicate symbols, and no function symbols or individual constants, is undecidable* [Kalmár, 1936: cf. Löwenheim, 1915]

For further negative results, see Church (1956, §47).

## 59 Odds and ends: 1. Logical validity and the empty domain. 2. Omega-inconsistency and omega-incompleteness. 3. Gödel's theorems. 4. The Axiom of Choice. 5. Recursively enumerable sets

*Logical validity and the empty domain*

We have been working with a technical notion of logical validity according to which a formula of Q is logically valid iff it is true for every interpretation with a *non-empty domain.* But (it might be argued) our technical definition does not catch satisfactorily a certain intuitive notion of logical validity according to which a formula is logically valid only if it is true (in an intuitive sense) for *all* interpretations (in an intuitive sense), including interpretations with an empty domain. It might be said: 'Standard logic certifies as logically valid (in its sense) formulas that are *not*, in an intuitive sense, logically valid: e.g. according to standard logic the formula

$$\bigvee xFx \vee \bigvee x \sim Fx$$

is logically valid. But is this formula logically valid, in an intuitive sense? Suppose nothing at all existed. Then, no matter what property was assigned to $F$, there would be nothing either to have or to lack the property. So for any interpretation (in an intuitive sense) with an empty domain the formula would be (in an intuitive sense) false.'

Recently there has been an increase of interest in versions of quantification theory that do not exclude the empty domain, and it is possible that inclusive quantification theory (as it is called) may in time to come supersede the standard theory covered in this book. But our main concern is with the standard theory, so we shall simply give a quotation and a reference. The quotation is from Quine (1953):

> An easy supplementary test enables us anyway, when we please, to decide whether a formula holds for the empty domain. We have only to mark the universal quantifications as true and the existential ones as false, and apply truth-table considerations.

The reference is to Meyer and Lambert (1967), which gives an axiom system for inclusive quantification theory and a short bibliography.[1]

*Omega-inconsistency. Omega-incompleteness*

Let S be a formal system that has the set of natural numbers as the domain of its intended interpretation. Then:

(1) S is said to be *ω-inconsistent* [omega-inconsistent] if there is a wff A(v) of S, with one free variable v, such that $\vdash_S A(\bar{n})$ for each natural number *n* and also $\vdash_S \sim\Lambda vA(v)$. S is said to be *ω-consistent* if there is no such wff.

If a first order theory is ω-consistent then it is consistent, but the converse is not true.

(2) S is said to be *ω-incomplete* [omega-incomplete] if there is a wff A(v) of S, with one free variable v, such that $\vdash_S A(\bar{n})$ for each natural number *n* but ΛvA(v) is not a theorem of S. S is said to be *ω-complete* if there is no such wff.

*Gödel's theorems*

1. *Gödel's completeness theorem*

In his paper 'Die Vollständigkeit der Axiome des logischen Funktionenkalküls'[2] [1930] Gödel proved the (semantic) completeness of a system of first order predicate logic that was essentially that of Whitehead and Russell's *Principia Mathematica.* In the course of the proof he showed also that every consistent set of wffs of the system is simultaneously satisfiable in a countable domain – a key result for later work.

2. *Gödel's (first) incompleteness theorem*

In his paper, 'Über formal unentscheidbare Sätze der Principia mathematica und verwandter Systeme I' [2] [1930, pub. 1931] Gödel proved that any formal system S satisfying the following three conditions:

1. S is ω-consistent
2. S has a recursively definable set of axioms and rules of inference
3. Every recursive relation is definable in S

is incomplete, in the sense that there is a closed wff of S of the

[1] A dissenting voice is that of the Kneales (1962, pp. 706–7), who argue that 'The reservation by which empty domains are excluded from consideration . . . is really no restriction at all because there can be no empty domains of individuals'.

[2] English translation in J. van Heijenoort, *From Frege to Gödel.*

form $\Lambda vFv$ such that neither $\vdash_S \Lambda vFv$ nor $\vdash_S \sim\Lambda vFv$, where on the intended interpretation F is assigned a recursively defined property of natural numbers.

This is Gödel's most celebrated theorem and the one usually meant when people speak of 'Gödel's Theorem'.

3. *Gödel's second incompleteness theorem* (*also called* '*Gödel's Second Theorem*')

In his 1931 paper Gödel treated mainly of a formal system that was essentially the result of adding Peano's axioms for arithmetic to the (higher order) logic of *Principia Mathematica.* He showed that a certain formula of the system that could be interpreted as asserting the consistency of the system was not a theorem of the system provided the system was consistent. This is Gödel's second incompleteness theorem. Gödel also made the more general claim that for this system (and for two others, one of them Zermelo–Fraenkel set theory) the consistency of the system is not provable in the system, provided the system is consistent. This claim, or some generalisation of it, may also be found described as 'Gödel's second incompleteness theorem'. However, Feferman (1956–7) showed, for a formal system of Peano's arithmetic, that there is a formula that in a sense expresses the consistency of the system and that is a theorem of the system. So care needs to be taken in stating exactly what has been proved in this area. Cf. also Mendelson, pp. 148–9, and Andrzej Mostowski (1964, pp. 23–6).

*The Axiom of Choice*

There are various ways of stating it (in fact there is a whole book, by H. and J. Rubin, devoted to *Equivalents of the Axiom of Choice*). A simple version is this:

For any set of non-empty disjoint sets [i.e. sets that have no members in common], there exists a set that has for its members exactly one member from each of those disjoint sets.

Another is:

For any set S of non-empty sets, there exists a function that assigns to each member X of S a member of X. [The function picks out for each set X (and there may be infinitely many Xs) a member of X: so infinitely many choices may be required. Hence the name of the axiom.] (This version does not require the members of S to be disjoint.)

The axiom is used not only in set theory but in the theory of transfinite numbers, in topology, and in measure theory. It is independent of the other axioms of set theory (provided those other axioms are consistent) [Cohen, 1963. Gödel in 1938 proved that it cannot be disproved from the other axioms].

'The axiom of choice is probably the most interesting and, in spite of its late appearance, the most discussed axiom of mathematics, second only to Euclid's axiom of parallels' (Fraenkel and Bar-Hillel, 1958, p. 47, which gives something of the history of the axiom). Besides the long discussion in Fraenkel and Bar-Hillel (pp. 44–80), there are relatively simple accounts of the axiom in Russell (1919, chap. xii) (where it is called 'the multiplicative axiom'), Rosenbloom (1950, pp. 146–51), and Suppes (1960, chap. 8). J. B. Rosser in his *Logic for Mathematicians*, pp. 510–12, raises the question 'How Indispensable Is the Axiom of Choice?' Non-Cantorian Set Theory [i.e. set theory with the negation of the axiom of choice in some form or other replacing the axiom of choice][1] is the subject of a clear and simple article by Paul Cohen and Reuben Hersh in the *Scientific American* [Cohen and Hersh, 1967). See also Cohen's book, *Set Theory and the Continuum Hypothesis.*

*Recursively enumerable sets*

A recursively enumerable set of natural numbers is, by Church's Thesis, the precisely defined equivalent of an effectively enumerable set of natural numbers. One way of defining an r.e. set is:

A set of natural numbers is *recursively enumerable* iff it is either empty or the range of a recursive function.

Intuitively (and assuming Church's Thesis) a set is recursively enumerable iff either it is empty or there is an effective method for generating its members one by one.

Emil Post, in a beautiful paper [Post, 1944], takes the notion of a recursively enumerable set as a starting point for a general

[1] The generalised continuum hypothesis implies the axiom of choice. So Non-Cantorian Set Theory has as a theorem the negation of the generalised continuum hypothesis.

theory of decision problems. Other fairly simple accounts of recursive function theory and decidability in which recursively enumerable sets are near the centre of the picture may be found in Davis (1966 and 1967) and in Rogers (1957). Cf. also Mostowski (1964, p. 39).

## Appendix 2

### *Synopsis of basic metatheoretical results for first order theories*

'Proved consistent [complete, decidable, etc.]' is short for 'Some formal system of . . . proved consistent [complete, decidable, etc.].'

### I. Logic

A. *Truth-functional propositional logic*

Proved consistent, semantically complete, syntactically complete, decidable. (Post, 1920)

Not negation-complete.

B. *First order monadic predicate logic*

Proved consistent, semantically complete, decidable (Löwenheim, 1915).

Not syntactically complete or negation-complete.

C. *First order predicate logic* (*with or without identity*)

Proved consistent (Hilbert and Ackermann, 1928), semantically complete (Gödel, 1930), undecidable (Church, 1936; assuming Church's Thesis).

Not syntactically complete or negation-complete.

The first complete set of axioms and rules for first order predicate logic was given by Gottlob Frege, in his *Begriffsschrift* (1878, pub. 1879).

## II. First Order Theories in General

A. The Löwenheim–Skolem Theorem: Every first order theory that has a model has a denumerable model (Löwenheim, 1915 – Skolem, 1919).

B. Every consistent first order theory has a model (Gödel, 1930).

## III. Mathematical Theories [1]

1. *Elementary number theory with addition but without multiplication*

Proved consistent, negation-complete, decidable. (Presburger, 1929)

2. *Elementary number theory with multiplication but without addition*

Proved consistent, negation-complete, decidable. (Skolem, 1930)

3. *Elementary algebra of real numbers, with addition and multiplication and every individual natural number, but without the general notion of a natural number*

Proved consistent, negation-complete, decidable. (Tarski, 1930, pub. 1948)

4. *Elementary geometry, with every individual natural number but without the general notion of a natural number*

Proved consistent, negation-complete, decidable. (Tarski, 1930, pub. 1948)

5. *Elementary number theory, with addition and multiplication, individual natural numbers and the general notion of a natural number*

Proved consistent (Gentzen, 1935, appealing to a principle [of transfinite induction] that belongs neither to first order predicate logic nor to elementary number theory itself, and so is at least as open to question as the theory whose consistency it is used to

[1] Cf. also Tarski, Mostowski and Robinson (1953). In the descriptions that follow, 'elementary' means that part of the full theory that can be expressed in a first order language without using concepts from set theory.

prove). Proved that any consistent formalisation of the theory with an effective method for telling whether or not a thing is a proof in the system is semantically incomplete (generalisation of Gödel, 1930, pub. 1931, which required $\omega$-consistency and a primitive recursive set of axioms. Rosser, 1936, showed that $\omega$-consistency could be replaced by consistency. The present formulation involves an appeal to Church's Thesis). Proved undecidable (Church, 1936; assuming Church's Thesis and the consistency of the system).

6. *Zermelo–Fraenkel set theory. Von Neumann–Bernays–Gödel set theory*

Not proved consistent. Proved if consistent then incomplete (Gödel, 1930, pub. 1931) and undecidable (corollary of Church, 1936).

## References

ACKERMANN, WILHELM (1896–1962)

1927 'Zum Hilbertschen Aufbau der reellen Zahlen', *Mathematische Annalen*, vol. 99 (1928) 118–33. Received 20 Jan 1927. Eng. trans. in van Heijenoort, pp. 495–507.

1928 'Über die Erfüllbarkeit gewisser Zählausdrücke', *Mathematische Annalen*, vol. 100 (1928) 638–49.

1954 *Solvable Cases of the Decision Problem* (North-Holland, Amsterdam, 1954).

ANDERSON, ALAN ROSS (b. 1925) and BELNAP, NUEL DINSMORE (b. 1930)

1959 'A simple treatment of truth functions', *Journal of Symbolic Logic*, vol. 24 (1959) 301–2.

BERNAYS, PAUL (b. 1888) and SCHÖNFINKEL, MOSES ( )

1927 'Zum Entscheidungsproblem der mathematischen Logik', *Mathematische Annalen*, vol. 99 (1928) 342–72. Received 24 Mar 1927.

CANTOR, GEORG (1845–1918)

1873 'Über eine Eigenschaft des Inbegriffes aller reellen algebraischen Zahlen', *Journal für die reine und angewandte Mathematik*, vol. 77 (1874) 258–62. Dated Berlin, 23 Dec 1873. Reprinted in Cantor (1932, pp. 115–18).

1877 'Ein Beitrag zur Mannigfaltigkeitslehre', *Journal für die reine und angewandte Mathematik*, vol. 84 (1878) 242–58. Dated Halle, 11 July 1877. Reprinted in Cantor (1932, pp. 119–33).

1895, 1897 'Beiträge zur Begründung der transfiniten Mengenlehre', *Mathematische Annalen*, vol. 46 (1895) 481–512; vol. 49 (1897) 207–46. Reprinted in Cantor (1932, pp. 282–351). Eng. trans. as *Contributions to the Founding of the Theory of Transfinite Numbers* (Open Court, Chicago, 1915; Dover, New York, 1952).

1932 *Georg Cantor Gesammelte Abhandlungen*, ed. E. Zermelo (Springer, Berlin, 1932). Reprinted by Georg Olms Verlagsbuchhandlung (Hildesheim, 1962).

Church, Alonzo (b. 1903)

1935*a* 'An unsolvable problem of elementary number theory. Preliminary report', *Bulletin of the American Mathematical Society* (May 1935), vol. 41, pp. 332–3. Received 22 Mar 1935.

1935*b* 'An unsolvable problem of elementary number theory', *American Journal of Mathematics*, vol. 58 (1936) 345–63. Presented to the Society 19 Apr 1935. Reprinted in Davis (1965, pp. 89–107).

1936 'A note on the Entscheidungsproblem', *Journal of Symbolic Logic*, vol. 1 (1936) 40–1 (correction pp. 101–102). Received 15 Apr 1936. Correction received 13 Aug 1936. Reprinted in Davis (1965, pp. 110–15).

1956 *Introduction to Mathematical Logic* (Princeton University Press, 1956).

Cohen, Paul Joseph (b. 1934)

1963 'The independence of the continuum hypothesis', *Proceedings of the National Academy of Sciences of the U.S.A.*, vol. 50 (1963) 1143–8; vol. 51 (1964) 105–10. Received 30 Sep 1963.

1966 *Set Theory and the Continuum Hypothesis* (W. A. Benjamin, Inc., New York, 1966).

Cohen, Paul Joseph (b. 1934) and Hersh, Reuben

1967 'Non-Cantorian Set Theory', *Scientific American*, vol. 217, no. 6 (Dec 1967) 104–6, 111–16. Reprinted in Morris Kline (ed.), *Mathematics in the Modern World* (W. H. Freeman and Company, San Francisco and London, 1968) pp. 212–20.

Davis, Martin (b. 1928)

1958 *Computability & Unsolvability* (McGraw-Hill, New York, 1958).

1965 (ed.), *The Undecidable* (Raven Press, Hewlett, N.Y., 1965).

1966 'Recursive Functions – An Introduction', in E. R. Caianiello (ed.), *Automata Theory* (Academic Press, New York and London, 1966) pp. 153–63.

1967 'Recursive Function Theory', in Paul Edwards (ed.), *Encyclopedia of Philosophy*, vol. 7 (Macmillan and Free Press, New York, 1967) 89–95.

DEDEKIND, RICHARD (1831–1916)

1887 *Was sind und was sollen die Zahlen?* (Vieweg, Brunswick, 1888). Preface dated 5 Oct 1887. Eng. trans. in Richard Dedekind, *Essays on the Theory of Numbers* (Open Court, Chicago, 1901; Dover, New York, 1963).

FEFERMAN, SOLOMON

1956–7 'Formal consistency proofs and interpretability of theories', doctoral dissertation, Berkeley, 1957. Abstracts in *Journal of Symbolic Logic*, vol. 22 (1957) 106–7 (received for publication 1956).

'Formal consistency proofs and interpretability of theories', *Summaries of talks presented at the Summer Institute for Symbolic Logic, Cornell University, 1957*, pp. 71–7.

'Arithmetization of metamathematics in a general setting', *Fundamenta Mathematicae*, vol. 49 (1960) 35–92. Received 10 Dec 1959.

FITCH, FREDERIC BRENTON (b. 1908)

1938 'The consistency of the ramified *Principia*', *Journal of Symbolic Logic*, vol. 3 (1938) 140–9.

FRAENKEL, ABRAHAM ADOLF (1891–1965)

1961 *Abstract Set Theory*, 2nd ed. (North-Holland, Amsterdam, 1961).

FRAENKEL, ABRAHAM ADOLF (1891–1965) and BAR-HILLEL, YEHOSHUA (b. 1915)

1958 *Foundations of Set Theory* (North-Holland, Amsterdam, 1958).

FREGE, FRIEDRICH LUDWIG GOTTLOB (1848–1925)

1878, pub. 1879 *Begriffsschrift* (Louis Nebert, Halle, 1879). Preface dated 18 Dec 1878. Eng. trans. in van Heijenoort, pp. 5–82.

GALILEO (1564–1642)

1638 *Discorsi e dimostrazioni matematiche, intorno a due nuove scienze* (The Elseviers, Leiden, 1638). Eng. trans. as *Dialogues concerning Two New Sciences* (Macmillan, New York, 1914; Dover, New York, 1960, where see 'First Day', pp. 31–2).

GENTZEN, GERHARD (1909–45)

1935 'Die Widerspruchfreiheit der reinen Zahlentheorie', *Mathematische Annalen*, vol. 112 (1936) 493–565.

Received 11 Aug 1935: some changes made in Feb 1936. Eng. trans. in *The Collected Papers of Gerhard Gentzen*, ed. M. E. Szabo (North-Holland, Amsterdam and London, 1969).

GÖDEL, KURT (b. 1906)

1930 'Über die Vollständigkeit des Logikkalküls', doctoral dissertation, University of Vienna. Degree granted 6 Feb 1930.
'Die Vollständigkeit der Axiome des logischen Funktionenkalküls', *Monatshefte für Mathematik und Physik*, vol. 37 (1930) 349–60. Eng. trans. in van Heijenoort, pp. 583–91.

1930, pub. 1931 'Über formal unentscheidbare Sätze der Principia mathematica und verwandter Systeme, I', *Monatshefte für Mathematik und Physik*, vol. 38 (1931) 173–98. Received for publication 17 Nov 1930. Abstract presented to the Vienna Academy of Sciences 23 Oct 1930. Eng. trans. in van Heijenoort, pp. 595–616.

1933 'Zum Entscheidungsproblem des logischen Funktionenkalküls', *Monatshefte für Mathematik und Physik*, vol. 40 (1933) 433–43.

1934 lectures *On Undecidable Propositions of Formal Mathematical Systems*, notes by S. C. Kleene and J. B. Rosser of lectures given at the Institute for Advanced Study, Princeton, during the spring of 1934. Printed with corrections and a postscript in Davis (1965, pp. 39–74).

1938 'The consistency of the axiom of choice and of the generalized continuum-hypothesis', *Proceedings of the National Academy of Sciences of the U.S.A.*, vol. 24 (1938) 556–7. Cf. also his *The Consistency of the Continuum Hypothesis* (Princeton University Press, 1940) (notes of lectures given in 1938).

GRZEGORCZYK, ANDRZEJ

1961 *Fonctions récursives* (Gauthier-Villars, Paris; Nauwelaerts, Louvain, 1961).

HARROP, RONALD

1964 'Some structure results for propositional calculi', *Journal of Symbolic Logic*, vol. 30 (1965) 271–92. Presented July 1964.

HENKIN, LEON (b. 1921)

1947 'The completeness of formal systems', doctoral dissertation, Princeton, 1947. Part published in 'The completeness of the first-order functional calculus', *Journal of Symbolic Logic*, vol. 14 (1949) 159–66 (reprinted in Hintikka, 1969, pp. 42–50). Another part in Henkin (1950) (reprinted in Hintikka, 1969, pp. 51–63).

1950 'Completeness in the theory of types', *Journal of Symbolic Logic*, vol. 15 (1950) 81–91. Received 11 Mar 1949. Reprinted in Hintikka (1969, pp. 51–63). Part of his doctoral dissertation (1947).

HERBRAND, JACQUES (1908–31)

1929, pub. 1930 'Recherches sur la théorie de la démonstration', doctoral dissertation for the University of Paris, published as *Travaux de la société des sciences et des lettres de Varsovie*, Classe III, Sciences mathématiques et physiques, Nr 33 (Warsaw, 1930) 128 pp. Dated 14 Apr 1929. Reprinted in his *Écrits logiques* (Presses Universitaires de France, Paris, 1968) pp. 35–153. For the proof of the Deduction Theorem see §2.4 of chap. 3 (p. 61 in the original, pp. 90–1 in the 1968 reprint). There is an Eng. trans. of chap. 5 in van Heijenoort, pp. 529–67.

HERMES, HANS (b. 1912)

1961 *Aufzählbarkeit, Entscheidbarkeit, Berechenbarkeit* (Springer, Berlin, 1961). Eng. trans. under the title *Enumerability, Decidability, Computability* (Springer, Berlin, 1965).

HILBERT, DAVID (1862–1943)

1904 'Über die Grundlagen der Logik und der Arithmetik', *Verhandlungen des Dritten Internationalen Mathematiker-Kongresses in Heidelberg vom 8. bis 13. August 1904* (Teubner, Leipzig, 1905). Eng. trans. in van Heijenoort, pp. 129–38.

1925 'Über das Unendliche' (address delivered in Münster, 4 June 1925), *Mathematische Annalen*, vol. 95 (1926) 161–90. Eng. trans. in van Heijenoort, pp. 369–92.

1927 'Die Grundlagen der Mathematik' (paper for Hamburg seminar, July 1927), *Abhandlungen aus dem mathematischen Seminar der Hamburgischen Universität*, vol. 6

(1928) 65–85. Reprinted in his *Die Grundlagen der Mathematik* (Teubner, Leipzig, 1928) and in his *Grundlagen der Geometrie* (Teubner, Leipzig and Berlin, 1930). Eng. trans. in van Heijenoort, pp. 464–79.

HILBERT, DAVID (1862–1943) and ACKERMANN, WILHELM (1896–1962)

1928 *Grundzüge der Theoretischen Logik* (Springer, Berlin, 1928). Eng. trans. of 2nd (1938) ed. under the title *Principles of Mathematical Logic* (Chelsea Publishing Co., New York, 1950).

HINTIKKA, KAARLO JAAKKO JUHANI (b. 1929)

1969 (ed.), *The Philosophy of Mathematics* (Oxford University Press, 1969).

HIŻ, HENRY (b. 1917)

1957 'Complete Sentential Calculus Admitting Extensions', *Summaries of talks presented at the Summer Institute for Symbolic Logic, Cornell University, 1957*, pp. 260–2. Abstract in *Notices of the American Mathematical Society*, vol. 5 (1958) 29. 'Extendible sentential calculus', *Journal of Symbolic Logic*, vol. 24 (1959) 193–202.

HOBBES, THOMAS (1588–1679)

1656 *Six Lessons to the Professors of the Mathematiques . . . in the University of Oxford* (Andrew Crook, London, 1656).

HUGHES, GEORGE E. and CRESSWELL, M. J.

1968 *An Introduction to Modal Logic* (Methuen, London, 1968).

KALMÁR, LÁSZLÓ (b. 1905)

1932 'Über die Erfüllbarkeit derjenigen Zählausdrücke, welche in der Normalform zwei benachbarte Allzeichen enthalten', *Mathematische Annalen*, vol. 108 (1933) 466–484. Received 3 Sep 1932.

1935 'Über die Axiomatisierbarkeit des Aussagenkalküls', *Acta Scientiarum Mathematicarum*, vol. 7 (1934–5) 222–243. Received 21 Mar 1935.

1936 'Zurückführung des Entscheidungsproblem auf den Fall von Formeln mit einer einzigen, binären, Funktionsvariablen', *Compositio Mathematica*, vol. 4 (1936–7) 137–144. Received 22 May 1936.

KLEENE, STEPHEN COLE (b. 1909)
1952 *Introduction to Metamathematics* (North-Holland, Amsterdam; P. Noordhoff, Groningen; Van Nostrand, New York, 1952, etc.).
1967 *Mathematical Logic* (Wiley, New York, 1967).

KNEALE, WILLIAM CALVERT (b. 1906) and KNEALE, MARTHA (b. 1909)
1962 *The Development of Logic* (Clarendon Press, Oxford, 1962).

LEBLANC, HUGUES (b. 1924)
1966 *Techniques of Deductive Inference* (Prentice-Hall, Inc., Englewood Cliffs, N.J., 1966).

LEMMON, EDWARD JOHN (1930–66)
1965 *Beginning Logic* (Nelson, London and Edinburgh, 1965).

LINDENBAUM, ADOLF (1904–41)
For Lindenbaum's Lemma, cf. Alfred Tarski, *Logic, Semantics, Metamathematics*, p. 98, Theorem 56.

ŁOS, JERZY
1951 'An algebraic proof of completeness for the two-valued propositional calculus', *Colloquium Mathematicum*, vol. 2 (1951) 236–40.

LÖWENHEIM, LEOPOLD (1878–*c.* 1940)
1915 'Über Möglichkeiten im Relativkalkül', *Mathematische Annalen*, vol. 76 (1915) 447–70. Eng. trans. in van Heijenoort, pp. 232–51.

ŁUKASIEWICZ, JAN (1878–1956)
1929 *Elementy logiki matematycznej*, lithographed (Warsaw, 1929). Eng. trans. of 2nd ed. (ed. J. Słupecki) under the title *Elements of Mathematical Logic* (Pergamon, Oxford, 1963).

MARGARIS, ANGELO
1967 *First Order Mathematical Logic* (Blaisdell Publishing Company (Ginn & Co.), Waltham, Mass., 1967).

MARKOV, ANDREI ANDREEVICH (the younger) (b. 1903)
1951 'The theory of algorithms' (in Russian), *Trudy Mathematicheskogo Instituta imeni V. A. Steklova*, vol. 38 (1951) 176–89.
1954 'The theory of algorithms' (in Russian), *Trudy Mathematicheskogo Instituta imeni V. A. Steklova*, vol. 42

(1954). Eng. trans. Israel Program for Scientific Translations, Jerusalem, 1961.

MATES, J. R. BENSON (b. 1919)

1953 *Stoic Logic* (University of California Press, Berkeley and Los Angeles, 1953).

1965 *Elementary Logic* (Oxford University Press, 1965).

MENDELSON, ELLIOTT

1964 *Introduction to Mathematical Logic* (Van Nostrand, Princeton, 1964).

MEYER, ROBERT K. and LAMBERT, KAREL (b. 1928)

1967 'Universally free logic and standard quantification theory', *Journal of Symbolic Logic*, vol. 33 (1968) 8–26. Received 9 Feb 1967.

MOSTOWSKI, ANDRZEJ

1964 'Thirty years of foundational studies', *Acta Philosophica Fennica* fasc. XVII (1965); also Blackwell, Oxford, 1966. Lectures given in 1964.

PEANO, GIUSEPPE (1858–1932)

1889 *Arithmetices principia, nova methodo exposita* (in Latin) (Turin, 1889) xvi +20 pp. Partial Eng. trans. in van Heijenoort, pp. 85–97.

PEIRCE, CHARLES SANDERS (1839–1914)

1867–1911 *Collected Papers*, ed. Charles Hartshorne and Paul Weiss, vols. III and IV (Harvard University Press, 1933).

PÉTER, RÓZSA (b. 1905)

1951 *Rekursive Funktionen* (Akadémiai Kiadó, Budapest, 1951). There is an Eng. trans. of the third, revised, edition under the title *Recursive Functions* (Academic Press, New York and London, 1967).

POST, EMIL LEON (1897–1954)

1920 Doctoral dissertation, Columbia University, 1920. Summary presented at a meeting of the American Mathematical Society, 24 Apr 1920. Published under the title 'Introduction to a general theory of elementary propositions', in *American Journal of Mathematics*, vol. 43 (1921) 163–85. Reprinted in van Heijenoort, pp. 265–83.

1936 'Finite combinatory processes. Formulation I', *Journal of Symbolic Logic*, vol. 1 (1936) 103–5. Received 7 Oct 1936. Reprinted in Davis (1965, pp. 289–91).

1942, pub. 1943 'Formal reductions of the general combinatorial decision problem', *American Journal of Mathematics*, vol. 65 (1943) 197–215. Received 14 Nov 1941. Revised 11 Apr 1942.

1944 'Recursively enumerable sets of positive integers and their decision problems', *Bulletin of the American Mathematical Society*, vol. 50 (1944) 284–316. Presented to the Society, 26 Feb 1944. Reprinted in Davis (1965, pp. 305–37).

PRESBURGER, MOJŻESZ

1929 'Über die Vollständigkeit eines gewissen Systems der Arithmetik ganzer Zahlen, in welchem die Addition als einzige Operation hervortritt', *Sprawozdanie z I Kongresu matematyków krajów slowiańskich, Warszawa, 1929* (Warsaw, 1930) pp. 92–101, 395.

QUINE, WILLARD VAN ORMAN (b. 1908)

1934–60 *Selected Logic Papers* (Random House, New York, 1966).

1937 'Completeness of the propositional calculus', *Journal of Symbolic Logic*, vol. 3 (1938) 37–40. Received 13 Sep 1937. Reprinted in his *Selected Logic Papers*, pp. 159–163.

1940 *Mathematical Logic* (Harvard University Press, 1940).

1953 'Quantification and the empty domain', *Journal of Symbolic Logic*, vol. 19 (1954) 177–9. Received 28 Sep 1953. Reprinted in his *Selected Logic Papers*, pp. 220–3.

ROBINSON, ABRAHAM

1966 *Non-standard Analysis* (North-Holland, Amsterdam, 1966).

ROBINSON, RAPHAEL MITCHEL

1950 'An essentially undecidable axiom system', *Proceedings of the International Congress of Mathematicians, Cambridge, Massachusetts, 1950* (American Mathematical Society, Providence, R.I., 1952), vol. I, pp. 729–730.

ROGERS, HARTLEY

1957 'The present theory of Turing machine computability', *Journal of the Society for Industrial and Applied Mathematics* [*S.I.A.M. Journal*], vol. 7 (1959) 114–30. Lecture given in 1957. Reprinted in Hintikka (1969, pp.130–46).

1967 *Theory of Recursive Functions and Effective Computability* (McGraw-Hill, New York, 1967).

ROSENBLOOM, PAUL C. (b. 1920)

1950 *Elements of Mathematical Logic* (Dover, New York, dated 1950, pub. 1951).

ROSSER, JOHN BARKLEY (b. 1907)

1936 'Extensions of some theorems of Gödel and Church', *Journal of Symbolic Logic*, vol. 1 (1936) 87–91. Presented to the Association for Symbolic Logic and the American Mathematical Society, 1 Sep 1936. Received for publication 8 Sep 1936. Reprinted in Davis (1965, pp. 231–5).

1953 *Logic for Mathematicians* (McGraw-Hill, New York, 1953).

RUBIN, HERMAN and RUBIN, JEAN E.

1963 *Equivalents of the Axiom of Choice* (North-Holland, Amsterdam, 1963).

RUSSELL, BERTRAND ARTHUR WILLIAM (1872–1970)

1919 *Introduction to Mathematical Philosophy* (Allen & Unwin, London, 1919).

SCHÜTTE, KURT (b. 1909)

1933 'Untersuchungen zum Entscheidungsproblem der mathematischen Logik', *Mathematische Annalen*, vol. 109 (1934) 572–603. Received 30 June 1933.

1954 'Ein System des verknüpfenden Schliessens', *Archiv für mathematische Logik und Grundlagenforschung*, vol. 2 (1956) 55–67. Received 9 Dec 1954.

SHEFFER, HENRY MAURICE (1883–1964)

1912 'A set of five independent postulates for Boolean algebras, with applications to logical constants', *Transactions of the American Mathematical Society*, vol. 14 (1913) 481–8. Presented to the Society 13 Dec 1912.

SHEPHERDSON, JOHN CEDRIC (b. 1926)

1967 'Algorithms, Turing Machines and Finite Automata; An Introductory Survey', in D. J. Stewart (ed.), *Automaton Theory and Learning Systems* (Academic Press, London and New York, 1967) pp. 1–22.

SHOENFIELD, JOSEPH R. (b. 1927)

1967 *Mathematical Logic* (Addison-Wesley, Reading, Mass., 1967).

SKOLEM, THORALF (1887–1963)
All the following papers by Skolem may be found, in German, in his *Selected Works in Logic*, ed. J. E. Fenstad (Universitetsforlaget, Oslo–Bergen–Tromsö, 1970).

1919 'Logisch-kombinatorische Untersuchungen über die Erfüllbarkeit oder Beweisbarkeit mathematischer Sätze nebst einem Theoreme über dichte Mengen', *Videnskapsselskapets Skrifter*, I. Matematisk-naturvidenskabelig Klasse, 1920, no. 4 (Kristiania, 1920) 36 pp. Presented 21 Mar 1919. Eng. trans. of §1 in van Heijenoort, pp. 254–63.

1922 'Einige Bemerkungen zur axiomatischen Begründung der Mengenlehre', *Conférences faites au cinquième congrès des mathématiciens scandinaves tenu à Helsingfors du 4 au 7 juillet 1922* (Helsingfors, 1923). First publication of the Skolem Paradox. Eng. trans. in van Heijenoort, pp. 291–301.

1923 'Begründung der elementaren Arithmetik durch die rekurrierende Denkweise ohne Anwendung scheinbarer Veränderlichen mit unendlichem Ausdehnungsbereich', *Videnskapsselskapets Skrifter*, I. Matematisk-naturvidenskabelig Klasse, 1923, no. 6 (Kristiania, 1923) 38 pp. Eng. trans. in van Heijenoort, pp. 303–33.

1928 'Über die mathematische Logik', *Norsk matematisk tidsskrift*, vol. 10 (1928) 125–42. Lecture given 22 Oct 1928. Eng. trans. in van Heijenoort, pp. 512–24.

1930 'Über einige Satzfunktionen in der Arithmetik' *Skrifter Utgitt av Det Norske Videnskaps-Akademi i Oslo*, I. Mat.-Naturv. Klasse, 1930, no. 7 (Oslo, 1931). Presented 12 Sep 1930.

1933 'Über die Unmöglichkeit einer vollständigen Charakterisierung der Zahlenreihe mittels eines endlichen Axiomensystems [D.h. wenn die Begriffe 'Menge' oder 'Aussagenfunktion' präzisiert werden]', *Norsk Matematisk Forenings Skrifter*, serie II, Nr. 10, pp. 73–82 (Oslo, 1933).

1934 'Über die Nicht-charakterisierbarkeit der Zahlenreihe mittels endlich oder abzählbar unendlich vieler Aussagen mit ausschliesslich Zahlenvariablen', *Fundamenta Mathematicae*, vol. 23 (1934) 150–61.

SMULLYAN, RAYMOND M.
1968 *First-Order Logic* (Springer-Verlag, Berlin, Heidelberg, New York, 1968).

STOLL, ROBERT R.
1963 *Set Theory and Logic* (W. H. Freeman, San Francisco, 1963).

STRAWSON, PETER FREDERICK (b. 1919)
1952 *Introduction to Logical Theory* (Methuen, London; Wiley, New York, 1952).

SUPPES, PATRICK (b. 1922)
1957 *Introduction to Logic* (Van Nostrand, Princeton, 1957).
1960 *Axiomatic Set Theory* (Van Nostrand, Princeton, 1960).

TARSKI, ALFRED (b. 1902)
1923–38 *Logic, Semantics, Metamathematics: Papers from 1923 to 1938*, trans. J. H. Woodger (Clarendon Press, Oxford, 1956). For the history of the Deduction Theorem, cf. fn. on p. 32.
1928 For the history of the Upward Löwenheim–Skolem Theorem, cf. the editor's remarks at the end of Skolem (1934) (where it is ascribed to Tarski, seminar at the University of Warsaw, 1927–8) and p. 94, fn. 8, of Tarski and Vaught (1956) (where Tarski gives the date as 1928). For proofs of the theorem, cf. Mendelson, p. 69, Margaris, p. 178.
1930, pub. 1948 *A decision method for elementary algebra and geometry*, prepared for publication by J. C. C. McKinsey (The RAND Corporation, Santa Monica, California, 1948) pp. iii + 60. Revised edition (University of California Press, Berkeley, 1951).

TARSKI, ALFRED (b. 1902), MOSTOWSKI, ANDRZEJ and ROBINSON, RAPHAEL MITCHEL
1953 *Undecidable Theories* (North-Holland, Amsterdam, 1953).

TARSKI, ALFRED (b. 1902) and VAUGHT, ROBERT L.
1956 'Arithmetical Extensions of Relational Systems', *Compositio Mathematica*, vol. 13 (1957) pp. 81–102. Received 16 Feb 1956.

TURING, ALAN MATHISON (1912–54)
1936 'On computable numbers, with an application to the

Entscheidungsproblem', *Proceedings of the London Mathematical Society*, series 2, vol. 42 (1936–7) 230–65 (corrections in vol. 43 (1937) 544–6). Received for publication 28 May 1936. Read 12 Nov 1936. Reprinted in Davis (1965, pp. 116–51) (corrections pp. 152–4).

VAN HEIJENOORT, JEAN

1967 *From Frege to Gödel: A Source Book in Mathematical Logic, 1879–1931* (Harvard University Press, 1967).

WHITEHEAD, ALFRED NORTH (1861–1947) and RUSSELL, BERTRAND (1872–1970)

1910–13 *Principia Mathematica*, 3 vols (Cambridge University Press, 1910–13; 2nd ed., with new introduction and appendices, 1925–7).

ŻYLIŃSKI, EUSTACHY

1924 'Some remarks concerning the theory of deduction', *Fundamenta Mathematicae*, vol. 7 (1925) 203–9. Dated May 1924.

# Index

DATE DUE

| | | | |
|---|---|---|---|
| | | | |
| | | | |
| | | | |
| | | | |
| | | | |
| | | | |
| | | | |
| | | | |
| | | | |
| | | | |